HACKING PRODUCT DESIGN

A GUIDE TO DESIGNING PRODUCTS FOR STARTUPS

Tony Jing

Apress®

Hacking Product Design: A Guide to Designing Products for Startups

Tony Jing
San Francisco, California, USA

ISBN-13 (pbk): 978-1-4842-3984-1   ISBN-13 (electronic): 978-1-4842-3985-8
https://doi.org/10.1007/978-1-4842-3985-8

Library of Congress Control Number: 2018959217

Managing Director, Apress Media LLC: Welmoed Spahr
Acquisitions Editor: Shiva Ramachandran
Development Editor: Laura Berendson
Coordinating Editor: Rita Fernando
Copy Editor: Michael G. Laraque

Cover designed by eStudioCalamar

Distributed to the book trade worldwide by Springer Science+Business Media New York, 233 Spring Street, 6th Floor, New York, NY 10013. Phone 1-800-SPRINGER, fax (201) 348-4505, e-mail orders-ny@springer-sbm.com, or visit www.springeronline.com. Apress Media, LLC is a California LLC and the sole member (owner) is Springer Science+Business Media Finance Inc (SSBM Finance Inc). SSBM Finance Inc is a **Delaware** corporation.

For information on translations, please e-mail rights@apress.com, or visit www.apress.com/rights-permissions.

Apress titles may be purchased in bulk for academic, corporate, or promotional use. eBook versions and licenses are also available for most titles. For more information, reference our Print and eBook Bulk Sales web page at www.apress.com/bulk-sales.

Any source code or other supplementary material referenced by the author in this book is available to readers on GitHub via the book's product page, located at www.apress.com/9781484239841. For more detailed information, please visit www.apress.com/source-code.

Printed on acid-free paper

For my parents,

*who taught me the value of hard work
and perseverance*

Contents

About the Author

Tony Jing works as a product designer at Uber Technologies Inc. Prior to that, he was a product designer at Inkling Systems Inc., a startup based in San Francisco. He also writes a popular blog on Medium, on topics such as design, prototyping, and technology in China.

Acknowledgments

I would like to thank Andrea Williamson, Andrew Hawryshkewich, and Russell Taylor for giving me the chance to discover design at university. I'd like to thank Peter Cho and Ryan Koziel for coaching me and giving me the opportunity to grow and flourish. I'd like to thank Chatree Campiranon, Zach Leach, Ed Lea, Albert Wang, Meagan Timney, and Elisha Ong for their mentorship and guidance. They have shown me what it means to do quality design work.

I'd also like to thank the team at Apress: Shiva Ramachandran, Rita Fernando, and Laura Berendson, for their support, persistence, and dedication in making this book a reality.

I must acknowledge the hundreds of writers and speakers who have shared their knowledge of the technology and design industry. Their insights, openness, and collaborative attitude have influenced me in profound ways.

Last, I'd like to say thanks to the readers of my Medium blog, whose feedback, and recommendations have motivated me to continue writing.

Introduction

Okay, you just started working as a product designer at a startup, and you ask yourself, What should I do? This book attempts to answer that question. Years ago, when I switched from graphic design to product design, I had the exact same question. I wished there was a book that covered all the soft skills related to designing products for startups.

This is why I wrote this book.

I want to help those who are, as I was, entering the industry for the first time. Whether you're working at a small five-person startup out of a coworking space or a fast-growing startup with hundreds of employees around the world, there are a number of generally accepted and shared practices about how technology products are designed and built.

This book is about the soft skills that a product designer should have in order to be successful working in a technology startup. It assumes that you already have some foundational hard skills in design work, such as visual design, interface design, information architecture, prototyping, copywriting, and motion design, or you will find ways to acquire those skills on your own. After all, there are plenty of books, articles, tutorials, and other resources out there teaching those skills. If you don't have them and don't plan on learning them, reading this book alone will not make you a product designer.

Hard skills are absolutely crucial to good design work. However, having only those skills will not release the full potential of design. Involvement with design remains at a production level, and without hard skills, one will not be able to rise to a position that can positively impact product strategy and a business's bottom line.

The aim of this book is to help designers unleash the full potential of design at a startup. It is divided into nine chapters, each covering a topic pertinent to the work of product design in startups. The chapters can be read together or separately.

Chapters 1 to 3 cover the foundational knowledge about what startups are, what mindsets designers should have when working in them, and what kinds of framing designers should have in order to solve problems. Chapters 4 to 6 are about ideas, specifically how to get them, how to collaborate with others to make the best out of them, and how to prioritize what to do in order to maximize the potential of those ideas. Chapters 7 to 9 showcase the relevant considerations, heuristics, and frameworks that can be applied to design for the right contexts.

Let's get started!

How Startups Work

Before diving into the details of how to design product for startups, it is important that I cover the fundamental concepts and terminologies used in the book. This chapter aims to introduce some of these concepts and terms, as well as the unique roles startups have played in the course of human history. It will also cover the basic realities of startups as businesses in the 21st century and introduce the concept of a product team—a team responsible for creating the products driving the successes and failures of startups.

Technology and Human History

The world, as we know it, has been shaped by the accumulation of human actions over thousands of years. While we pay considerable attention to the impact of individual actions on historical events, the impact of technological progression on human civilization can often be overlooked. However, if we simply stop to look around, we easily discover evidence of such impacts. Take, for example, the names we give to prehistoric periods—*Stone Age, Bronze Age,* and *Iron Age*—which indicate the three technological materials on which humans then relied.

In the past three hundred years, the unprecedented pace of technological change has led to the coinage of new terms, such as *Industrial Age, Atomic Age, Space Age,* and *Information Age,* as illustrated in Figure 1-1. Some of these terms

© Tony Jing 2018
T. Jing, *Hacking Product Design,* https://doi.org/10.1007/978-1-4842-3985-8_1

may seem antiquated today, but they designate the significant transformations that human societies have experienced, which may represent the most consequential three centuries in human history.

Figure 1-1. Atomic Age, Space Age, Information Age

What's especially noteworthy is that in the last 30 years, the pace of that change is increasing. The rise of microcomputing, and then the Internet, has transformed the world in ways beyond the wildest imagination of the public a few decades earlier. This is the world we live in today.

How Do Startups Fit into This?

In part, the Industrial Revolution has led to today's economic systems. Most of the world's societies have embraced market economies. Despite the differences in state-sanctioned policies among countries, individuals by and large have the freedom and means to participate in buying and selling goods and services. Startup companies operate within this global reality.

The dictionary definition of a startup is simply a newly established business. From Berlin to Bogota, Lagos to London, Shanghai to San Francisco, most people have the option to establish a business without prohibitive restrictions.

The motivations to start a business can be incredibly nuanced, but three basic categories exist: wealth accumulation, personal achievement, and helping others. Before starting businesses, individuals often assess economic factors around them, as well as their own abilities to estimate the possibilities of success. Then they go about registering the business with the local government.

Starting up could be as simple as that. In fact, every single company in the world, from a mom-and-pop shop to a global business conglomerate, began with an idea and an assessment of factors.

What makes startups interesting in today's world is the combination of the freedom and ease to establish new businesses and the breakneck advancements in technologies related to computing and the Internet. This combination has allowed some companies to grow very rapidly.

Table 1-1 illustrates perfectly the impact of startups and their technologies on human progress. Among the top-ten companies in 1987, IBM was the only one that operated in the field of computing. In 2017, the number rose to five (shown in boldface). In fact, three of the top-ten companies (Alphabet, Amazon.com, and Facebook) weren't even founded in 1987. The reason for this growth is simple—the products and services produced by these companies were deemed so valuable that people clamored to use them. These companies grew from startups in Silicon Valley garages to huge global enterprises in a matter of one or two decades.

Table 1-1. Fortune 500's Top-Ten Companies in 1987[1] vs. the Top-Ten Companies Ranked by Market Capitalization in January 2017[2]

1987	2017
General Motors	**Apple Inc.**
Exxon Mobil	**Alphabet Inc.**
Ford Motor	**Microsoft**
IBM	Amazon.com
Mobil	Berkshire Hathaway
General Electric	ExxonMobil
AT&T	Johnson & Johnson
Texaco	**Facebook**
DuPont	JPMorgan Chase
Chevron Texaco	Wells Fargo

---

[1]FORTUNE 500 Archive List, http://archive.fortune.com/magazines/fortune/fortune500_archive/full/1987/.
[2]Financial Times, "FT Global 500," https://markets.ft.com/data/dataarchive/ajax/fetchreport?reportCode=GMKT&documentKey=688_GMKT_170401.

How Does Design Fit into This?

Quite simply put, design is the conception and planning of the tools humans use and interact with. From the moment prehistoric humans first picked up a wedge-shaped stone and used it as a chopper, design has become a part of our experience (Figure 1-2). The early humans who planned and conceived of using that stone were the first designers. Over time, tools diversified and became much more complex.

Figure 1-2. Design is as old as humanity

Now tools are ubiquitous in everyday life, in every corner of human existence. They affect the clothes we wear, the buildings we live in, the roads we walk on. In fact, every single human-made object is the result of design. Tools have become systems, platforms, networks, and nonphysical things. Beyond utility and function, aesthetics have also become a big part of design.

The technological sea change in the last three decades has led to the need for a new type of designer. Products today are no longer just physical; they are made of bits as well as atoms. Soon, most of the world's communication will occur digitally. Business will be conducted almost exclusively on the Internet. In this reality, the new type of designer needs not only to work within this digital realm but also manage a high level of complexity in these systems.

Designers must also be well versed in the high-level knowledge of computing and the Internet, as well as the specific industry context in which their businesses operate.

The Modern Startup

The term *startup* has been co-opted by the media, entrepreneurs, and technology investors to mean a newly formed business in industries related to technology and the Internet that is intended to grow very quickly. They aim to do so by developing innovative products and/or services that often replace, reinvent ("disrupt"), or improve upon traditional ways of life and commerce as they existed before the explosion of personal computing and the Internet.

Paul Graham, a renowned computer programmer, entrepreneur, and venture capitalist, transcribed his thoughts on startups. To him, the most important characteristic of a startup is growth, in that the startup is designed specifically to grow rapidly. His point is that out of the millions of new companies created every year, most are traditional service-based businesses, such as barber shops or restaurants. These are not startups, in that they are not intended to grow fast—in revenue, profits, number of customers, etc. The fundamental difference between a startup and traditional business, according to Graham, is that not only do startups "make something lots of people want," which traditional businesses do as well, they also "reach and serve all those people" in ways traditional services-based businesses (e.g., a barbershop) cannot.[3] This new reach is mostly owing to the Internet, which did not exist 30 years ago.

The trait of innovative reach can be seen in most startup companies. Unlike your local community center, for example, Facebook is a hub that connects all of the world. Unlike your local radio station, Spotify is a music-streaming service for the entire globe. Unlike your local public storage, Dropbox is a storage service accessible from anywhere with wi-fi.

Startups Are Fickle

The upside of startups is that this new reach allows for the potential to grow and scale up a business quickly, if the product or service is done right. The downside of this reach is that it now has to compete with the entire Internet, which potentially has hundreds of thousands of businesses potentially doing the same thing.

---

[3]Paul Graham, "Startup = Growth," www.paulgraham.com/growth.html, September 2012.

In fact, by their nature, startups are fickle. Most of them fail—an overwhelming majority of them within the first few years. The unique advantage a startup has is that the rapid progression of technology has yet to show signs of a significant slowdown. This constant progress allows doors to be opened that were unimaginable a few years previous. Steve Jobs and Steve Wozniak wanted to assemble a personal computer in 1975, back when no one wanted one. Larry Page and Sergey Brin wanted to create a search engine when most people didn't know what a search engine was. The key to really hitting it big is to be able to peer beyond the horizon and anticipate something big that's about to happen.

However, peering into the future is almost akin to trying to predict the future—most people get it wrong. This is why initiating a startup can be incredibly difficult. If you're joining the industry because you think it will guarantee you wealth, you couldn't be more wrong. For every Google or Facebook, there are tens of thousands of failed companies that were often started by people just as smart and driven as the Google and Facebook founders. The journey of starting a technology company can be fraught with hardships, surprises, and failures. However, the companies that battle through these challenges and continue to grow over time are the ones that change the world.

Startup Terminology

Following are the terms that people working in startups are quite familiar with. They will be referred to later in this book.

> **Agile:** A method of developing software that emphasizes adaptability, collaboration, and cross-functional teams
>
> **Angel investor:** An individual who provides the initial capital for a business to start, in exchange for equity/ shares in the company
>
> **B2B:** A business model in which the target customers are other businesses, rather than individual persons. The opposite of a B2B company is a *consumer* company.
>
> **Bootstrap:** To bootstrap is to start and grow a company without outside investment.
>
> **Equity:** Ownership shares in a company, often used to exchange for capital
>
> **Growth hacking:** A buzzword for marketing with data-driven methodologies

Incubator: Organizations that help to develop early-stage startups by providing initial funding, industry insights, connections with investors, and access to a network of entrepreneurs, usually for a small amount of equity in the company

IPO: An initial public offering. This allows a company to be listed on a public stock exchange. Typically, after an IPO, a company is no longer considered a startup. Investors often hope to make a profit through an IPO of a startup.

MVP: Minimum viable product. This refers to the simplest version of the product, in terms of features, that could satisfy customer needs, built with the goal of soliciting feedback, which enables quick and iterative product development.

Pivot: A change in business strategy, often done to promote better profitability and a more sustainable business model

Product: A term that refers to the product *and* service that a company sells

SaaS: Software as a service. This is a business model in which software is provided as an on-demand service via the Internet. Most SaaS companies are also B2B companies.

Scope: The information that is explicitly expressed before working on a new project. Specifically, *project scope* refers to the work that must be completed in order to finish the project. Product scope refers to the planned features and functionalities of a new product.

Unit economics: The net revenue and cost number, expressed on a per-unit basis (per sale, per usage, etc.). It is usually used to measure the business viability and sustainability of a startup.

Valuation: A company's worth in monetary terms. This is usually determined by assessing the company's current capital structure, revenue, and future potential.

Venture capital: The money provided to startups that allows them to grow. Capitalization is usually done in series, typically after the initial capital provided by the angel investor or the incubator and after the startup has demonstrated growth traction.

The Product Team

The nucleus group of a startup is its product team. While there are many functions of a business, the product team is key to the success of the startup, because it directly impacts whether and how people will use the product.

The size of a product team can vary between three to fifteen people. This number greatly depends on the size of the startup, the industry it's in, and its specific stage of growth. At a minimum, the product team should consist of a product manager, a product designer, and a few engineers. If possible, data scientists, user researchers, and product marketers also should be included. Depending on the nature of the business, a project manager, an operational manager, and client success managers could also be embedded with the product team.

The tasks of the product team are simple: (a) figure out what to build and how to build it; (b) build an MVP and validate it by testing, then learn from the tests; (c) improve upon the MVP, based on what's learned, then test again.

Unlike the executive decisions made by the company leadership, the product team is responsible for decisions that actually execute the roadmap, by figuring out what details and features should be included, so that the customers' problems can be solved, and the company's vision can become a reality. Of course, if the company is a small five-person team, this separation between company and product team won't exist. However, what is constant in startups large and small is the iterative product development cycle of researching, hypothesizing, building, testing, and iterating.

Product Manager

A product manager (PM) has two key responsibilities. First, the product manager should work with the company leadership team to figure out what should be included in the roadmap of a product. In essence, the product manager should capture opportunities where the company mission can be realized in the form of a product. *Ideas are cheap.* They can come from anyone and anywhere. It is up to the product manager to come up with a framework to score, filter, and rank problems worth solving.

Second, the product manager is responsible for determining what goes into a product and what stays out, in other words, defining the product features and functionalities, by embodying the customer's pain points and the company's unique advantages. It is his/her job to work with customers directly, to figure out what the goals are, or with researchers and salespeople, who are proxies for the customer's voice. Product management entails a considerable amount of collaboration with other teams, conducting primary and secondary research, then coming up with a set of heuristics to evaluate customer problems that are worth solving. After this, the product manager works with designers and engineers to hypothesize product details. These two tasks of figuring out what problems to solve for and what solutions to hypothesize are central to the product manager's job.

Product Designer

Product designers are responsible for the core need of the customer being addressed by the products they design. They work with product managers to figure out a product's features and functionalities. Product designers are also responsible for the experience of a product. Unlike user-experience designers, a product designer cannot be blind to aspects of the product outside of the user experience. In other words, if a product doesn't address the users' needs, their experience doesn't matter. The product designer should address the fundamentals first.

After the features and functionalities of the product have been scoped out, it is the designer's job to turn these into concrete plans and artifacts that engineers can refer to to build out the features. Specifically, the designer is responsible for creating a vision of how a product could be used successfully in real life, architecting the experience in broad strokes first, then filling in the details of how the interface would work and what specific interactions models should exist. The goal should be to make something people want. Much of this book will cover the product designer's process and goals and how to achieve those goals within a startup product team.

Project Manager and Engineering

The project manager is the one who keeps track of the schedules of engineering tasks and deliverables. Engineering leads or managers sometimes take on this role. Large projects often have a dedicated project manager. Engineers are the ones building the product. Unlike physical product, digital ones don't require a manufacturing process, thus the engineers are the ones making the products.

It is important to note that engineering's role could vary greatly for each product, depending on the industry and its scope. However, in almost every case, engineers should be involved early in the problem definition phase of the project, to ensure the proper capture of the technological constraints and landscape.

Data Science, User Research, and Product Marketing

Data scientists are often embedded in a product team, to conduct quantitative research and to gain insights into how the product is performing. These insights are internalized by the team and used to make better product decisions. This role is especially important for products with large-scale user-base or usage patterns. User research tackles the qualitative side of insight gathering. It works with real users in a small sample size, to gather information on their perceptions, beliefs, and underlying behaviors.

Product marketing is responsible for telling the story of the product to an external audience. It also involves collaborating with sales and marketing for the startup as a whole. The product marketing role is often overlooked, but it is very important to a product's success, especially for those for which good public perception is key.

Summary

Product designers must be keenly aware of the world that their businesses operate within. They must be able to speak the startup language and understand the impact their decisions have on the business. Furthermore, they must closely collaborate with other team members, in order to create successful products. I'll delve into each of these topics in the next chapters.

Design Is a Mindset

Products exist to help people. We do things better, faster, or easier because of the products we use. The best products are so good at this that, eventually, they become natural extensions of our actions, ingrained in our everyday lives, and absorbed into the human experience. These products allow us to become better versions of ourselves. They do so by surfacing the values, virtues, and aspirations that are latent within us. This is why design is so important, because it complements the human condition.

Empathy

What do putting push/pull labels on doors, hanging sunglasses on shirts, and wrapping colored stickers around keys have in common? The answer is that they are all small acts that people do to improve their environments. Applying an empathetic approach can allow designers to uncover opportunities to gather insights for new and novel designs that benefit people from similar small acts.

More often than not, the challenges people face aren't things that product designers face personally. The reason for this is simple: we don't represent everyone in the world. Our experience, assumptions, and knowledge about the world don't lend themselves to every scenario people face. This begs the question: How can we design for humans if we don't know the problems and challenges that humans face?

© Tony Jing 2018

T. Jing, *Hacking Product Design*, https://doi.org/10.1007/978-1-4842-3985-8_2

The answer is that designers should conduct research to understand people. Especially, being empathetic while conducting research can help to reach that goal. Empathy, in the context of design, is the intentional setting aside of our knowledge, experience, opinions, and, sometimes, worldview, to understand the perspectives, experiences, goals, motivations, aspirations, and expectations of the people for whom we're designing.

This does not mean the wholesale abandonment of ourselves. It means that we must let go of our egos and biases, in order to absorb and meaningfully understand the situations of other people. It means a heightened sense of awareness of people other than us.

Product design is about solving problems and aspiring for the betterment of humanity. To do so, being empathetic is almost always the first step. To get the best out of the design process, product designers should systematically employ a few methods to empathize with the people they are trying to help.

Egos and Assumptions

Letting go of our egos and assumptions is perhaps the hardest task. This is not because we're stubborn or difficult but, rather, because we're often unaware of our own assumptions and ego. The first step in letting go starts with being honest and open, honest in terms of being aware of the moments when we defend or justify our own biases when faced with new situations, and open in terms that we're able to absorb new ideas without judging them.

The second step to letting go of your ego is to realize the need to win and to be right. Design is not about proving you're right or wrong. Design is about serving the needs of people and the world as a whole. It is important to consciously uncouple yourself from those needs.

Listen and Observe

Listening to people talk about their challenges is a cheap and fast way to quickly gather basic facts and information about a design problem. It also allows you to build rapport with the target audience, so that they'll become more natural and unreserved when you observe their actions, behaviors, and reactions.

The best way to understand people is to experience life in their shoes. This can be partially achieved by going to their locations to observe and experience their activities and environments. By placing ourselves in their physical environment and observing what they do, how they do it, and the challenges they face along the way, we can relate to their experiences much more accurately and innately.

When listening is combined with observing and embodying, designers can understand the motivations, needs, and goals behind people's actions and behaviors in a much more deep and profound way.

Body Language and What's Not Said

Designers must study body language, signals, facial expressions, voice intonations, and the positive and negative connotations associated with them. This is a skill that comes with practice. If mastered, it can help to uncover the goals and motivations of the target audience effectively and efficiently. Here are some tips to get started:

- Become adept at detecting the subtle nuances in speech, for example, changes of tone, points of pauses.
- Listen for what has been redacted or rephrased.

Often, people can only articulate part of the picture, as they might not have the full view themselves. In addition, people's memories are notoriously faulty. That's not to say that we shouldn't trust people's descriptions of their own experiences. We should listen and take notes but also back up those notes with our own observations.

Accounting for a lack of clarity and visibility and faulty memories, people may still not be able to convey all the required details. They might be burdened by fear, distrust, or embarrassment. As designers, we must hone our intuition, train our emotional sensitivity, to be able to uncover those critical details of people's reflections of their motivations, needs, and goals, without intruding into their personal lives or making people feel uncomfortable.

Example of Lack of Empathy

In 2012, Google launched an augmented reality wearable computer, called Project Glass, to much fanfare. Google Glass, as it's known to the public, was introduced by Google cofounder Sergey Brin live on stage in Moscone Center, San Francisco. The product was delivered to Brin by a team of skydivers and BMX athletes, who live streamed their descent from a plane in the sky onto the roof of the Moscone Center then all the way onto the stage.

People genuinely felt excited about the product at the time. However, launching a consumer product to the public is very different from launching a research prototype to a handful of beta-testers. Two years later, Google scrapped its consumer program for Glass.

While there were many reasons why Glass didn't live up to its hype, two main ones stood out. First, people didn't find many compelling use cases for it. You could take photos, shoot videos, send messages, and get directions on

Glass, but the experience wasn't a whole lot better than on the phone. While it might be fantastic for streaming first-person point-of-view videos, most people didn't have a need to share videos from a first-person perspective all the time.

Second, as well as being voice-controlled, Google Glass was designed to be worn on the face. Having a device that can record without other people's knowledge made those around Google Glass wearers uncomfortable. Interacting with a voice-controlled device in public was still considered a socially awkward behavior back then. *MIT Technology Review* put it best,

> No one could understand why you'd want to have that thing on your face, in the way of normal social interaction.[1]

You might be wondering how a multi-billion company can make a mistake like this? Perhaps hubris over having achieved a technological feat led to a lackluster effort in researching empathy about why the product needed to exist in the first place. Ultimately, the reason didn't really matter. What matters is that the Glass team lacked the empathy to address difficulties related to human interaction, privacy concerns, and social expectations. Ultimately, the idea of putting a computer on one's face, as illustrated in Figure 2-1, was rejected by the consumer.

Figure 2-1. Google Glass required users to wear a computer on their face

[1]Rachel Metz, "Google Glass Is Dead; Long Live Smart Glasses," *MIT Technology Review*, November 26, 2014, www.technologyreview.com/s/532691/google-glass-is-dead-long-live-smart-glasses/.

This example highlights the precise blind spot that product designers must address, along with the help of product managers and researchers. It is the product team's responsibility to make sure that products exist for a reason and that the team must empathetically understand the goals and motivation of the end users of its product.

Curiosity and an Open Mind

An essential quality for any good designer is having a keen eye for improvements. Where others put up with problems, product designers should look to fix them. They should notice the poor experiences, however subtle, that have long been accepted as the norm.

To do that, one must be willing to exert a lot of energy to learn, discover, and experiment and to question the assumptions and consider the obvious. Elon Musk, the renowned entrepreneur behind PayPal, Tesla, and SpaceX, explained this as "reasoning from first principles," or formulating a complex idea on the basis of fundamental truths. The example he gave was the basis for founding SpaceX, his private aerospace manufacturing and space transport company. He asked, "If the raw materials that make up a rocket—the metals, electronics, computers, and fuel—is only X, why is the cost of a rocket 100 times X?"[2] When Musk realized that there had to be some (if not many) inefficiencies in the process, he saw a tremendous opportunity. He developed this opportunity further, and, over time, SpaceX became the first aerospace manufacturing and space transportation company to feature reusable rockets, as shown in Figure 2-2.

---

[2]Elon Musk, "The First Principles Method Explained by Elon Musk," YouTube video, www.youtube.com/watch?v=NV3sB1RgzTI, 2:48, posted by innomind, December 4, 2013.

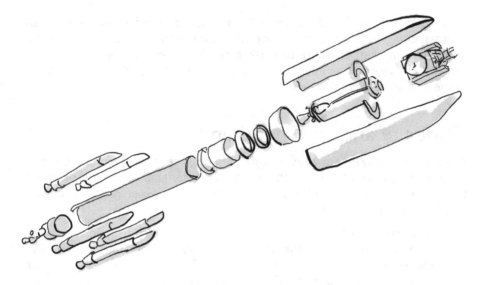

Figure 2-2. The many parts that make up a SpaceX rocket

Tony Fadell, the inventor and entrepreneur who cocreated the iPod at Apple and later founded Nest Labs, described the same method as viewing the world from a child's perspective and then asking the "why can't" questions. He cites as an example how his young son, when asked to go outside to check if there was mail in the box, responded by asking the question, "Why can't the mailbox check for mail itself and tell us?"[3]

Why can't it? That's a great question. The answer is that going outside to the box is how checking for mail always has been done, but when we think closely about that question, we realize that the status quo isn't necessarily the most convenient approach to doing things. This brings us to the second part of the product design mindset, which is to not accept the status quo for its own sake but to really think hard about whether it makes sense. Through this lens, to design means to know when a reexamination of what we know is required and to reimagine what's possible.

Have Virtues

Design has the power to shape our thoughts and behavior. The objects and environments around people unconsciously shape their feelings and perspectives. While keeping a curious and open mind allows the impossible to become possible and eventually the reality, the first step of design starts

---

[3]"The First Secret of Great Design," YouTube video, www.youtube.com/watch?v= 9u0MectkCCs, 16:41, posted by TED, June 3, 2015.

with defining our own personal values and virtues. It starts with asking the question, What kind of world do we want to live in?

Irene Au, the former head of design at Google and a design partner at Khosla Ventures, summarized this concept very succinctly: "Design is the culmination of intention, values, and principles manifested in tangible form and passed on to another."[4] It is important for product designers to clearly identify the virtues, values, and intentions they want to convey in their products. In other words, our personal values and mission should be aligned with those of our team, which, in turn, allows the team to create products that convey those values and virtues.

Never Stop Learning

Being a designer is about nonstop learning. Technology changes, and people change. As a result, products have to adapt to the changing times. New features must be added; outdated ones should be removed.

Once in a while, a new wave of innovative technology comes along that reshapes the entire landscape of human activities. Such times create new opportunities for new categories of products. Entire societies adopt completely new ways of life. We live in such a moment.

To design for today's world means to constantly learn and adapt. Designers not only have to keep up with the pace of innovation, they must lead it, along with engineers. Only when technology is built with people in mind can it truly benefit humans and push humanity forward.

Our job as designers is to connect humanity to things—to make things work for us, and not the other way around. To do this, designers must learn about new technologies. They must seek out new ways of getting humans to make sense of technology. This means paying attention to industry standards and shifts in those standards, testing and evaluating emergent technologies.

When it comes to the craft of designing for this new digital way of life, the dust has not yet settled on the standards of design tools and platforms. This means that designers must also stay flexible and absorb the changing tools and software involved in the creation of digital designs. Learning in design is a huge topic, which will be explored more in depth in Chapter 3.

---

[4]Irene Au, "Design and the Self," Medium.com, https://medium.com/design-your-life/design-and-the-self-a5670a000fee, August 10, 2016.

How to Solve a Problem

Beyond having empathy and an open, learning-oriented mindset, the core fundamental skill that a designer must possess is problem solving.

Frame, Then Reframe

A joke I heard early in my design education goes like this: How many designers does it take to change a light bulb? Answer: Does it have to be a light bulb? The point of the joke is that designers are often the ones asking the most obvious questions.

This notion of asking obvious questions ties back to Tony Fadell's advice about seeing the world through children's eyes. At its root, this question is about challenging the status quo for the sake of creating something new.

In fact, challenging the status quo is simply reframing the problem—challenging the prior assumptions by looking beyond their face value. What is obvious doesn't necessarily equate to relevancy or correctness. As designers, we should lead the team to think this way, in tandem with the product manager. More concretely, each new project should start by answering the following two questions:

- What do we stand for?

- What are the problems (in this space/industry)?

Answering these two questions before defining the product scope can help to guide product and design direction to a likelier chance of hitting the target of providing value.

An example of this is a project involving Chicago's troubled buildings that a student from the Institute of Design in that city worked on. When the project started, the initial problem that the city of Chicago gave the student was to make the process of tearing down vacant and abandoned buildings more efficient.

As Professor Jeremy Alexis of the Institute of Design explains,[5] after students collected data and conducted research, it became clear that the premise and assumptions made by the city were misaligned with the problem. The trend of buildings eventually becoming vacant and abandoned over time was actually not irreversible. It was a condition that could be changed and corrected, a fact that became the focus of the project. This is a prime example of reframing.

---

[5]"What Is Problem Framing in Design?" Vimeo video, https://vimeo.com/6180364, 2:43, posted by IIT Institute of Design, August 19, 2009.

After an insight, it became clear to both the students and the city that preventing buildings from becoming vacant and abandoned should be the focus of the project. This shift opened up the solutions space that targeted the root of the problem and, at the same time, created ideas that were much less expensive and faster to implement. Ideas such as using transit ads, utility bill inserts, community hearings, and presentations to engage and empower the community were brought forward, then they were reinforced using sticky notes, banners, and doorknob hangers to inform the community of abandoned building demolitions. Over time, the root cause of vacant and abandoned buildings was identified and ideas to solve it more effectively and efficiently were generated.

Get to the Constraints

In a sense, constraints make design happen. Innovation doesn't exist in a reality where everything is possible.

Here's why. Let's say you want to create a machine that enables instantaneous travel—something that gets you from A to B within a snap of the fingers. This is not feasible, at least not with our current technology. In this case, what you're proposing is nothing more than science fiction. It is not real. What this shows is that without constraints, there is no product design. Without constraints, what you're creating is fantasy. For product design to occur, what you come up with must eventually be made real. So, get to the edge of your constraints. Understand why the constraints are there. Know what's possible. This is why designers should be familiar with cutting-edges technologies.

Having Goals

In the context of designing and building a product, problem solving must be guided by specific goals.

Useful, Usable, Desirable

In general, it can be said that people want things that are useful, usable, and desirable—things that are simple to use yet add a lot of value to life. Design is an act balancing these three criteria. To start out, optimize for usefulness and usability. A good product must first be useful and usable.

Take the example of a pair of scissors, specifically those typically made for arts and crafts. If this pair of scissors isn't sharp enough, they won't perform the basic task of cutting paper. They aren't useful, in this case. However, a sharp pair of scissors might still not be useful, if the handles are not ergonomic.

When you cut with them for a while, your hands might become sore. If this is the case, the product is not usable.

Airbnb is a lot more useful than Craigslist when it comes to listing your home online to attract potential renters. Airbnb is more useful, because it is a service dedicated toward that goal. It has a significantly more focused user experience and a refined set of flows that is optimized for the purpose of listing and renting vacation homes.

Craigslist, one could argue, might be slightly more usable for the same purpose. Craigslist requires no background checks and no photos of your place when listing. All you need is an e-mail address. The difference in usefulness and usability between the two products is significant, but it is not astronomical.

The difference in growth and revenue between these two companies, however, is astronomical. Why is this so? One word: desirability. This is the other critical third dimension, other than usefulness and usability. It is an often overlooked one that truly separates great products from mediocre ones.

This is why products that are both useful and usable but lack desirability often don't sell. Steve Jobs explained this in plain language in 1997, when describing the products Apple should be building to turn around the company: "All we have to do is hold this up and say, 'Do you want this?'"

Ten years later, Jobs demonstrated this litmus test when he presented the first iPhone. The response in the subsequent years to the question Do you want an iPhone? has been a resounding yes. People literally line up around street corners to be among the first to purchase the latest model of the phone year after year.

The original iPhone, as shown in Figure 2-3, met all three goals discussed. It was useful, in that it was a phone, an Internet browser, and a music player rolled into one device—something that the market had never seen before. It was usable, in that even toddlers could pick it up and perform basic functions with it. It was desirable, in that people couldn't wait to own one, as it looked and felt good, and soon became a status symbol.

Figure 2-3. The first iPhone, unveiled in 2007

Of course, outside of consumer products, the nuances are much more refined, so holding up the product and asking people whether they want it wouldn't necessarily work, but the gist of designing for desirability remains the goal.

Viable, Feasible, Desirable

A bigger picture beyond the useful, usable, desirable framework is the viability, feasibility, and desirability framework. Quite simply put, this framework goes beyond the usefulness and usability of a product. It lays the three key questions that any team should ask before starting a new project.

1. Is it good for business?

2. Is it doable?

3. Is it something people want?

Just because a design is useful, usable, and desirable doesn't mean that it can be produced, manufactured, or engineered. Just because it can be engineered, it doesn't mean that the product is sustainable or makes economic sense. Product designers must not only act as the stewards of product usability and desirability but also safeguard against the lack of viability and feasibility.

The Ultimate Chart

A hidden cost of our designs is the effect they have on society and the natural environment. While it is often nearly impossible to foresee the future, designers should do their best at anticipating possible detriments to the world beyond their business and customers.

Figure 2-4 introduces a Venn diagram of goals that designers should adopt in evaluating a new project.

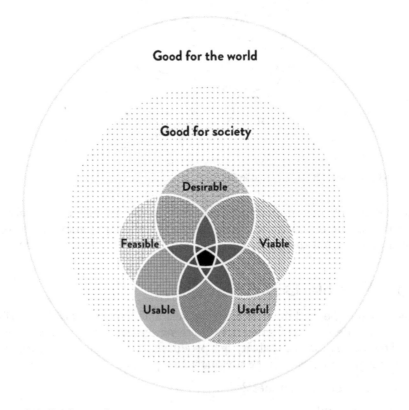

Figure 2-4. Product goals

Define the Product

It is especially important for startups to define the product(s) that it wants to create. Product designers should play a crucial role in defining the vision of the product, in accordance with the previously mentioned values, virtues, and mission, both theirs and the company's.

Some have referred to this process as developing a *minimum viable personality*.[6] Others have called this the creation of a *minimum desirable product*.[7] I think the core idea of all this minimum XY jargon is the same. It revolves around one question: How can we create a satisfying product experience?

Stand for Something

There are three steps to getting this right. First, we have to stand for something. We must have a vision of the future, an answer to the question of why we're doing what we do—and a vision of the world that we hope our product will help to bring about. In doing so, we're offering our customers a reason to stay, even if the product is not good yet. Tesla's roadster is a perfect example of this. It was a bold statement, a stake in the ground claiming the kind of future that Tesla hoped to stand for. Even though the MVP (minimum viable product) was full of issues, consumers stuck around.

Have a Guiding Principle

Second, the product should have a guiding principle, a gestalt, a core experience or technology that is ten times better than what is out there.[8] When Google's search and the iPhone's touch screen were first introduced, they were ten times better than what was out there. There has to be something that stands out so much (in a good way) that customers can easily evangelize or defend your product. Just having a vision is not enough. Your product must have at least one single experience that is ten times better than the competition's.

---

[6]Fred Wilson, "Minimum Viable Personality," AVC blog, http://avc.com/2011/09/minimum-viable-personality/, September 2011.

[7]Andrew Chen, "Minimum Desirable Product," http://andrewchen.co/minimum-desirable-product, December 7, 2009.

[8]Mark Suster, "Your Product Needs to Be 10x Better Than the Competition to Win. Here's Why," BothSidesOfTheTable.com, https://bothsidesofthetable.com/your-product-needs-to-be-10x-better-than-the-competition-to-win-here-s-why-6168bab60de7#.x0irrj37i, March 11, 2011.

Get Enough Details Right

Third, the product has to get enough details right. Not all the details, but enough have to be right. Lines, layout, and colors matter. Words, images, and illustrations matter. Interactions and animations matter. To determine the details that you really have to get right, think about what the core paths and features of the product are and whether those experiences are delightful. In the case of the original iPhone, there were many things that weren't good (battery life, camera, price, etc.), but the core experience of listening to music, browsing the Web, scrolling through a tracklist, and the cover flow were absolutely delightful. Steve Jobs even took a few seconds to showcase the "rubber banding" effect of a list bouncing delightfully when it reached the top,[9] as shown in Figure 2-5.

Figure 2-5. Steve Jobs demoing the original iPhone's "rubber banding" effect

Summary

Design requires a deep understanding of human nature: what makes us tick, what we trust, what we'll do repeatedly, and what is negligible. Of course, in practice, product designers must consider more than that. They exist in a

---

[9]Steve Jobs, "Steve Jobs iPhone 2007 Presentation (HD)," YouTube video, www.youtube.com/watch?v=vN4U5Fqr0dQ&feature=youtu.be&t=16m40s, 51:18, posted by Jonathan Turetta, March 13, 2013.

constant state of negotiation between entrepreneurs and engineers, between lofty visions and realistic capabilities. They must compromise, prioritize, and deliberate between what needs to be built and what can be. But while product designers are situated between external tensions, true product design begins in the mindset of the designer.

It is a mindset that puts humans first, not machines, objects, or systems. It is a mindset that does away with dogma, breaks meritless conventions, and tosses out untrue assumptions. It's a mindset that blends the idealism of entrepreneurship, the rigors of engineering, the emotiveness of art, and the understanding of human nature, to create things that truly *help and benefit* human beings.

Practice, Tasks, and Experiences

An often-overlooked fact about design is that most people already know how to do it. If you have sketched out an idea of something that you want to make, whether it's a toy, a piece of furniture, or an article of clothing, you have designed. At its essence, design is no more than the sketching out of ideas that can be turned into actual objects and systems. From that perspective, anyone who has ever created anything is a designer.

The first part of this chapter expands on this idea of everyone being a potential designer. It explains how that concept extends to designing for startups and emphasizes the importance of mastering one's craft as an essential prerequisite for doing good design work.

The second part of the chapter is about what product design really is. Product design is more than simply creating a physical or digital object. It is about the tasks people accomplish with the help of the products that are designed and built and the experiences they have while doing so.

A Craftsperson's Mindset

I choose a block of marble and chop off whatever I don't need.

This quote is often attributed to the French sculptor Auguste Rodin. This simple statement reflects years of craft and the accumulation of skill. A lot of effort and dedication are required before becoming a master craftsperson or artist.

© Tony Jing 2018

T. Jing, *Hacking Product Design*, https://doi.org/10.1007/978-1-4842-3985-8_3

We can all visualize figures, animals, sceneries in our heads, similar to how Rodin visualized a man in deep thought sitting on a rock (Figure 3-1). But chipping away pieces of marble to reveal realistic figures, animals, and sceneries is a much more difficult skill—to the extent that only a handful of people could do it as well as Rodin. That is why he is considered a master, one of the greatest sculptors of all time.

Figure 3-1. Rodin's The Thinker

A large part of Rodin's achievement was owing to his mastery of the mallet and the chisel. Not only could he visualize the final artwork inside the marble, he could command his toolset in such a way as to make that man a reality. That ability is what separates Rodin from the rest of us.

One needs to know the exact angles and the force to apply, in order to ensure that each strike against the marble is executed with purpose. The strikes must be applied in a sequence, removing layers of rock in a way that doesn't damage the final piece. This single practice of carving rock with precision takes years to master, notwithstanding the other skills required to become a master sculptor.

For Rodin, the mallet and the chisel were among the primary tools for creating a marble sculpture. The mastery of tools is what allows a craftsperson to become one with them, to the point where the tools come to feel like hands—a natural, intuitive part of a craftsperson's physical being. When that has been achieved, the craftsperson can refocus on solving design problems, without having to worry about sloppy execution of his/her vision.

This is the core idea behind becoming good designers. A mastery of tools is a must in the field of design. The reason is that, for the most part, product design is about communicating ideas, whether to a team of thousands of people or just to oneself. For someone designing and building a simple wooden stool to be used in his or her backyard, perhaps a simple sketch on paper suffices. However, in order to build a large seafaring container ship, teams of designers will have to communicate and collaborate with hundreds or thousands of people, using tools such as drawings, write-ups, videos, or 3D models.

In order to design great work, we must master the necessary tools. Our mastery must be that the tools become second nature. The exact tools may change and be updated, but what remains the same is the need for mastery.

Practice Makes Perfect

An often-cited parable in the design industry comes from the book *Art and Fear*,[1] in which the authors recount a story of a ceramics class.

> The ceramics teacher announced on opening day that he was dividing the class into two groups. All those on the left side of the studio, he said, would be graded solely on the quantity of work they produced, all those on the right solely on its quality. His procedure was simple: on the final day of class he would bring in his bathroom scales and weigh the work of the "quantity" group: fifty pounds of pots rated an "A", forty pounds a "B", and so on. Those being graded on "quality", however, needed to produce only one pot—albeit a perfect one—to get an "A". Well, came grading time and a curious fact emerged: the works of highest quality were all produced by the group being graded for quantity. It seems that while the "quantity" group was busily churning out piles of work—and learning from their mistakes—the "quality" group had sat theorizing about perfection, and in the end had little more to show for their efforts than grandiose theories and a pile of dead clay.

---

[1]David Bayles and Ted Orland, *Art & Fear* (Saint Paul, MN: Image Continuum Press, 1993), p. 29.

Practice does make perfect. Nested inside this cliché is the truth that good work is almost never the result of sheer genius. In fact, the notion of a "flash of genius" is almost always the result of continual practice and dedication. Behind every display of seemingly natural and innate talent is the prerequisite of thousands of iterations of practice and trial. Child prodigies are often simply kids who began learning and practicing certain skills at a young age and continued to hone their craft as they grew older. Bill Gates started programming when he was 13 years old, at a time when most 13-year-olds hadn't even seen a computer. His school happened to be one of the first to buy a computer for its students. Mozart learned to play the piano from his father at age four. What separated Bill Gates and Mozart from other people is that they never stopped perfecting their crafts (Figure 3-2).

Figure 3-2. Bill Gates and Mozart

The concept that talent is a manifestation of dedication and practice has been well-documented in recent years in books such as *Outliers: The Story of Success*, by Malcolm Gladwell (Little, Brown, 2008); *Talent Is Overrated*, by Geoff Colvin (Portfolio/Penguin, 2008); and *Grit: The Story of Passion and Perseverance*, by Angela Duckworth (Scribner, 2016). Collectively, the books give us the anecdotal and academic evidence of people achieving incredible feats through dedicated and focused practice, which often lasts years, if not decades.

Deliberate practice is a core idea of these books. First pioneered by K. Anders Ericsson, professor of psychology at Florida State University, deliberate practice is a concept by which an individual makes a conscious and focused practice of specific skillsets in a particular domain over long periods of time. More simply put, deliberate practice is repeatedly challenging oneself to accomplish tasks that are slightly more difficult than they would be otherwise. Throughout the process, practitioners remain completely focused and engaged with the task of practice and improvement. After many repeated cycles of deliberate practice, a person's overall skill level in the chosen domain increases drastically.

This idea of deliberate practice applies to every field of work. To become good at designing products, one must design a lot of products. Behind this tautology is a fundamental requirement: one must be willing to do a lot of work. Especially if you're new to the industry, the best way to start is to work on all the projects you can find, especially the ones in the fields you want to work in. Beyond that, the next thing is to master the tools, by using them and learning about them as much as possible.

Sketching

Ultimately, unlike engineering, product design is not responsible for the building, implementing, or manufacturing of the final design. Design and engineering should work closely together, but, ultimately, design doesn't go the final mile in the journey of bringing a product to life.

Instead, design communicates the ideas about the product through demonstrations, planning, and modeling. Designers are responsible for communicating what the solution is and how that solution can come about. The common skill found among all designers is the ability to sketch ideas quickly. This is the foundational skill to being a designer of any kind.

The goal of sketching is not to create a beautiful drawing, it is to communicate ideas in the most efficient way possible. Therefore, sketching is a necessary skill. Quite simply put, this is a skill that most of us already have. If you know how to draw circles, rectangles, and lines, you know how to sketch. Capture the basic form of something by using the simplest shapes.

At its core, design can be as simple as sketching ideas on the back of napkins. Throughout the design career progression, sketching is perhaps the single most universal skillset. For seasoned architects such as Frank Gehry, sketching is a significant aspect of his work, as his team is focused on turning those designs into reality.

Don't edit your sketches. Sketches are most commonly done in the beginning stages of product design. This means that they are supposed to be high-level ideas rather than nitty-gritty details. They are meant to express quick, unfiltered thoughts that can be quickly iterated later.

Tasks and Experiences

Product design is about solving problems through the creation of objects, experiences, systems, and networks. However, it can be said that everyone is a designer, in that we're all problem solvers. The following section covers some of the common methods and components that complement the problem-solving aspect of product design.

Jobs to Be Done

A good way to frame a design problem is to use the jobs-to-be-done model. Alan Klement,[2] a writer and business consultant, coined the term "Jobs to be Done" from insights gleaned on the product team at Intercom.[3] The core idea of this model can be summarized in the following formula:

> *When ____, I want to ____, so I can ____.*

An example of this could be, When I get up in the morning, I want to know if it will rain today, so I can bring an umbrella if it does. The "when" describes the situation aced. "I want" describes the motivations, the needs, and the desired goals that relate to the new situation. "I can" describes the expected outcomes or the real goals, which might be just below the surface and not directly revealed.

Using this model brings clarity to the process of determining what the outcome of your design problem should be. From that outcome, the product and its inner workings can be reverse-engineered step by step.

This idea of viewing business or design problems as "jobs" was first introduced by Harvard Business School professor Clay Christensen, who used this model to explain, in a talk he gave in 2007, how McDonald's increased the sales of its milkshakes.[4] The milkshake sales were improved based on feedback from customers.

This is when Christensen reminded his audience that there was a job that needed to be done when customers bought McDonald's. The key was to find that job and do it better. One of Christensen's colleagues stood in a McDonald's for 18 hours in one day, observing each transaction that involved milkshakes, noting down who bought them, what time it was, whether customers were in a group or by themselves, and what other items these customers bought.

A pattern quickly emerged. It turns out that more than half of the milkshakes sold that day were in the early morning. The people who bought the shakes were almost always by themselves. Also, they only bought the milkshakes, nothing else. Afterward, they would not drink the milkshake in the restaurant but drove away immediately after receiving their purchase.

---

[2]Alan Klement, "Replacing the User Story with the Job Story," JTBD.info, https://jtbd.info/replacing-the-user-story-with-the-job-story-af7cdee10c27, November 12, 2013.

[3]Paul Adams, "The Dribbblisation of Design," Inside Intercom, www.intercom.com/blog/the-dribbblisation-of-design/, May 21, 2018.

[4]Clay Christensen, "Jobs-to-Be-Done—Prof. Clayton Christensen," YouTube video, www.youtube.com/watch?v=Q63PZR7mG70, 7:56, posted by Strategsys, May 13, 2017.

To figure out what the job was, the team returned and questioned the same customers. They were asked what jobs they did that make them come to McDonald's at 6:30 a.m. to get a milkshake. The customers were bewildered by the unusual question, but after some explanation and rephrasing by Christensen's team, the patrons revealed that their jobs were the same. They all had a long, boring drive to work. The milkshake was something to keep them engaged, to not fall asleep, and to get something in their stomach before a 10 a.m. break time, when they could eat again. Of course, one of their hands had to be on the steering wheel, so the milkshake worked perfectly, as it only required one hand to consume (along with the straw), whereas such foods as bananas or donuts required either two hands, made things mucky, or were not filling enough. In addition, milkshakes, unlike coffee, felt like a real food, and even if the container was tipped over, it wouldn't spill. On top of that, it took more than 20 minutes to consume a milkshake, keeping the customer engaged for much of the commute.

McDonald's took those insights and resolved to improve the job solution, not just the milkshake. The first thing they did was to make the milkshake thicker, to last longer during the commute. Second, they added chunks of fruit to the milkshake, so that the commute became less predictable and more interesting. Last, they moved the dispensing machine from behind the cashier's counter to the front of the counter and installed an automated card-swipe payment system, so that customers could skip lines and self-serve, saving more time from their commute. By making these changes, the sales of milkshakes were increased by four times.

The jobs-to-be-done model also can be viewed from a goals and experience perspective, that is, the product should always address the true user goals in a way that provides an ideal experience.

Goals and Experiences

Products aren't simply the pixels, or materials, of the things we create. They aren't simply inner workings of a system. Instead, we're building ways people can accomplish their goals, by helping them (and us) do what we can't or have a hard time doing.

This concept of helping people accomplish their goals is about making products useful. Beyond that, as alluded to in Chapter 2, a seemingly useful product that is utterly unusable ceases to have any value. A useful and usable product that is totally undesirable also won't be used very often.

Making something useful, usable, and desirable is a core component of product design. The process of doing so is often referred to as *experience design*.

Experience design, or user experience (UX) design, entails all facets of how a system responds and interacts with a user of that system or systems. Good

UX design makes the user feel at home, in control, and engaged. Bad UX design makes people frustrated, confused, and anxious. Following is an example of how experience design, when done right, can help to transform an industry.

In 2001, Apple launched retail stores for the first time in its history, to the ridicule of much of the computer industry. Headlines such as "Sorry, Steve: Here's Why Apple Stores Won't Work"[5] dominated tech media. People scoffed at the idea, because companies such as Dell and HP struggled to create successful retail stores that sold computers. However, as seen in a 2001 promotional video of the Apple Store, founder and CEO Steve Jobs laid out a vision for these brick-and-mortar entities.

> *People don't just want to buy personal computers anymore, they want to know what they can do with them. We are going to show people exactly that.*
>
> —Steve Jobs[6]

What the team at Apple realized back then was that while many of their competitors focused on product and sales, few were focused on the user goal of understanding, learning, and using computers for specific use cases, as well as the experience of actually using the computers. People may have heard about getting a computer, but they had to discover what that meant. With that insight, Apple was able to go against the industry wisdom and open physical locations where people could try Apple computers for themselves. This opened up the entire market.

Rather than sticking to the tech-savvy crowd that most other companies targeted, Apple was able to open doors for those who were new to the idea of using a computer. As Steve Jobs explains in the store intro video,

> *Every product we make is in the first 25% of the store. As you see up on the ceiling, we've even labeled the section—home, music, genius kids, and pro, movies, photos, etc., on this side…Every single one of the computers is connected to the Internet… you can experience it for yourself.*

> *… Most of the products are running self-running demos… the solutions we've chosen to feature now are music, movies, photos, and kids…*

---

[5]Cliff Edwards, "Commentary: Sorry, Steve: Here's Why Apple Stores Won't Work," Bloomberg.com, www.bloomberg.com/news/articles/2001-05-20/commentary-sorry-steve-heres-why-apple-stores-wont-work, May 21, 2002.
[6]Steve Jobs, "Steve Jobs Introduces the Apple Store (2001)," YouTube video, www.youtube.com/watch?v=xLTNfIaL5YI, 4:14, posted by vintagemacmuseum, May 19, 2011.

> *Wouldn't it be great if you went to buy a computer, or after you've bought a computer, if you have questions, you can ask a genius? This is called the Genius Bar... There will be somebody here who can do service right in the store, who can answer any questions you got...and if that person doesn't know the answer, they have a hotline to call us in Cupertino in Apple headquarters, where we have somebody who does.*[7]

In hindsight, this strategy couldn't be simpler. Instead of plastering the store with detailed specifications of product hardware and software, the retail store revolved around showcasing what a user can do with a computer and resolving the challenges a user might come across, before, during, and after a purchase. Apple created a unique experience, where shopping for a computer became approachable and even joyful.

What Apple did is the essence of experience design, as examined through the lens of a startup. Apple addressed the fundamental needs and challenges that a particular set of users faced, by creating solutions that effectively addressed those needs and communicating those solutions to these users.

Let's break this down. Back in the late 1990s, people used computers for a variety of reasons. Some used them for business—writing documents and creating presentations and spreadsheets. Others used them for gaming. Some used computers to create software programs for other computers. Apple examined the landscape and understood that competition for these advanced users was already very intense, and Apple's existing product line simply wasn't compelling enough to capture these savvy users.

Therefore, Apple avoided competing for those users altogether. Instead, they expanded the market for personal computers by targeting new users, who could benefit from a computer's capability in delivering entertainment, education, and information-sharing. At the same time, Apple's line of higher-end computers was specifically tailored for creatives—digital professionals whose jobs required special software and whose needs were different than those of the office worker, the gamer, and the programmer.

This was the right thing to do. The industry was booming back then. The public was curious about personal computers. It wanted to find out more. However, when potential customers walked into typical computer retailers, their experience was often disappointing. This was because the existing marketing model showcased computers that were largely tailored to savvy users. Naturally, the curious layperson soon became alienated by the cold, hard technical specifications and found computers to be difficult things for which they had neither the knowledge nor time.

---

[7]Ibid.

This was precisely the insight that Apple understood before everyone else. One key takeaway here is that industry-changing product design starts with seeing major sea changes in the technological landscape in a timely manner. As a side note, this focus on strategy is typically not considered to be within the scope of experience design, but as discussed in Chapter 2, business viability and technical feasibility are both core to the product design discipline, especially in the fickle world of startup, in which a startup's life could hinge on proving its business model before the next round of venture capital funding.

This product strategy enabled Apple stores' experience design to be successful. However, while the product development teams executed on this strategy, the biggest hurdle Apple faced was communicating its newfound insight to its new potential audience. Putting Apple products alongside other specs-focused products in big-box computer retailers was the wrong thing to do, because these retailers don't have the same agency to tell the story of Apple. Therefore, the need to build retail stores became apparent.

The design of the Apple store empathetically served these new users' needs, allowing them to see for themselves what computers can do for them, as Steve Jobs explained in the promo video previously mentioned. The store's clean, elegant, and yet often daring architecture and interior design further enhanced the experience of those who visited and helped to create a level of user satisfaction unheard of in the computer industry. However, this wouldn't be possible without properly informing the target audience of the solutions tailored to their needs. The combined strategy of hosting events, inviting media, and creating straightforward explanatory videos were a big part of this communication. The biggest and most direct form of communicating the solutions was stores' locations. Apple outbid other retail companies for these expensive heavy foot-traffic locations in major metropolitan areas. This allowed passersby to see for themselves what the stores were about.

Altogether, the combination of product strategy, positioning, execution on user experience, design, and marketing made the new stores a tremendous success. This new way of conducting business not only reshaped the computing industry, it shaped the entire retail industry and beyond. Fast forward to today: the Apple store has become a cornerstone of Apple's slew of software and hardware businesses.

To startups designing a new product, the key takeaways from Apple's lessons are

1. Find an underserved user goal or need.

2. Empathetically develop ways to improve user experience while fulfilling those goals.

3. Clearly and directly communicate those improvements to users.

Summary

To ensure long-term success of the products they build, individual designers must embrace the mindset of the craftsperson. First and foremost, master the tools of the trade, and then continuously practice the craft of designing in a deliberate and focused way.

Meaningful experiences are one of the core outputs of good product design. This is the baseline requirement of a product designer, especially in a startup building digital products. However, to build a truly great product, as Apple has shown, business strategy and communication methods must be well thought out. Product designers must be co-owners in that process of shaping business strategy and communication methods, especially in a startup setting.

Evaluating and Informing Ideas

Before We Start Designing

In this chapter, I will cover the ways in which product ideas can be formed, explored, and evaluated for startups. These are the essential skills that separate startup product designers from UX designers at large companies or graphic designers whose job scope does not include many of the business and personal considerations.

Ideas are a dime a dozen. It only is through the process of design, testing, feedback, and iteration that a startup can evaluate the business impact of an idea. It is much cheaper and faster to design something and build a prototype to test it out than to build the real thing.

In this chapter, I will discuss the process that product designers can adopt to help startups test and evaluate ideas. I will showcase the common types of product ideas that startups develop and explain how research and data can be used to guide idea validation.

© Tony Jing 2018
T. Jing, *Hacking Product Design*, https://doi.org/10.1007/978-1-4842-3985-8_4

Changing Times

There was a meme circulating around the Internet in mid-2016[1]:

1998:

- *Don't get in strangers' cars*
- *Don't meet ppl from internet*

2016:

- *Literally summon strangers from internet to get in their car*

This meme (expressed visually in Figure 4-1) speaks volumes about the kind of changes society has seen. Ideas that were once laughed upon or simply deemed unfathomable are now the norm, because of changes in new technologies. However, the process of using technology to create such changes requires near constant evaluation and insights gathered from people's interaction with those products. This chapter is about how product designers can apply their skills and perspectives within a product team, to help bring about those insights.

Figure 4-1. Summoning strangers from the Internet to get in their cars

Unlike graphic designers, product designers who work in startups are responsible for evaluating the ideas behind products and businesses. Startups, by their definition, are fickle, and ideas are cheap. As alluded to in the story about the ceramics class in Chapter 3, success is almost always the result of iteration through feedback. What makes a product stand out is the process of constant iteration and improvement. Seemingly far-fetched ideas eventually provide billions of dollars' worth of utility to people around the world.

---

[1]Carol Nichols, Twitter post, https://twitter.com/Carols10cents/status/749109677431021568, July 1, 2016, 10:17 p.m.

Amazon, Google, Facebook, Uber, Airbnb, all grew exponentially by following a similar approach.

Startup product designers have a special opportunity to make this process a key win for their team. This is because they straddle the space between product definition and user experience definition. They are the link ensuring that what the team builds both solves the right problem and is created in a way that is usable and feels right.

Product designers can do this by helping to define the problem being solved, specifically what jobs are to be done, and then leading the rounds of design iteration and design execution. Therefore, it is especially important that product designers know how to generate ideas, evaluate them, and then turn their insights into actionable items for themselves and for the team.

Getting Ideas

The idea for a technology business or startup can be generalized roughly into three categories. They are

1. Simplify

2. Me too

3. Virtualize

These categories can be overlain, remixed, or even combined, but nearly all startups and product ideas fall into them. For example, the various online tax-filing software virtualizes the task of visiting a tax accountant in person to do one's taxes. In 2016, Facebook launched a "me too" product, called Chatbots, by expanding the WeChat public accounts concept across the world. The various ride-sharing apps, such as Uber and Lyft, however, simplified the process of calling a car to two steps: open up an app, then press a button.

However, it is not essential that a startup enter an industry knowing the type of idea it should concentrate on. Success and failure can result from any type of idea, and in every market. What is important is that the idea must satisfy a need or a want in a way that is a lot better than the competitors'. In other words, the product must do the job users hire it to do better than what's already on the market.

The renowned Internet entrepreneur Bill Gross drives this point home in an interview with the seasoned angel investor Mark Suster.[2] Gross said that

---

[2]Mark Suster, "Your Product Needs to Be 10x Better Than the Competition to Win. Here's Why," Both Sides of the Table, https://bothsidesofthetable.com/your-product-needs-to-be-10x-better-than-the-competition-to-win-here-s-why-6168bab60de7, March 11, 2011.

a (new) product needs to be ten times (10x) better than its competitors' in order to be a success. While exactly what is ten times better is hard to quantify, aiming for a ten times North Star ensures that your product has a chance of standing out. The team, however, should not fixate on quantifying the improvements. Rather, it should focus on creating a product that is a lot better.

So, how do you get to a 10x product?

Generating Startup Ideas

There isn't one single way to come up with good ideas for products or businesses. Often, the best ideas come from problems we experience. This is not to say that you cannot solve a problem you don't personally experience. The key is to really understand the problem, and the industry. Only after that understanding is acquired can a product truly have the potential to thrive and become viable. Only when a product is viable can a great user experience be created to enhance it.

By becoming empathetic and adopting the techniques mentioned in Chapter 2, product designers can lead the effort to understand the problem space. Then, as Clay Christensen highlighted in Chapter 3, the jobs that must be done can be identified. Analyze the jobs in your market or field with product managers. Assist your project manager in getting as much clarity on that as possible.

If your team is stuck in a rut trying to create a new product, stop and look around first. Try to identify any aspects of life that can be improved. Ask yourself: What are the scenarios and situations we're trying to solve for? What are the jobs that people need done? Are there things that can be simplified? Are there products and services that can be brought to another market? Keep looking and never lose sight of the goal: building a product that does the jobs that people need done that is 10x better than what's incumbent.

Asking these questions will help you to frame your perspectives in thinking about the world for what it could be. Paul Graham,[3] the renowned entrepreneur and investor, summarized this sentiment best.

> Live in the future, then build what's missing.

This quote is especially true if you're building an innovative and technology-driven product. Graham's statement captures the unique responsibility and opportunity that a product designer has in a startup. Like product managers, product designers should also innately understand the problems that could be

---

[3]Paul Graham, "How to Get Startup Ideas," Paulgraham.com, www.paulgraham.com/startupideas.html, November 2012.

eradicated by the team. Unlike product managers, however, product designers bear the unique responsibility of making sure the solution being proposed, tested, and built is truly created for jobs to be done, provides the ideal human experience (as illustrated by the Apple store example in Chapter 3), and not the reverse—shoehorning the human experience into an existing technology (as illustrated by the Google Glass example in Chapter 2).

The key here is differentiating between the problems people *say* they have and the problems that they'll *pay* to address, even incompletely. Sometimes, people will casually say that they want something, but when you ask them to pay for it, they back out. These are the types of problems you want to avoid. Paul Graham used the analogy of wells, to illustrate the difference between the two types of startup ideas. The first is one that is broad and shallow, as in it being something that is sort of a nice-to-have for a lot of people. The latter, however, is narrow and deep, meaning that a few people want that product a lot (Figure 4-2). Microsoft, Apple, Yahoo, Google, and Facebook all started this way, as niche products that a relatively small group of people wanted immensely.

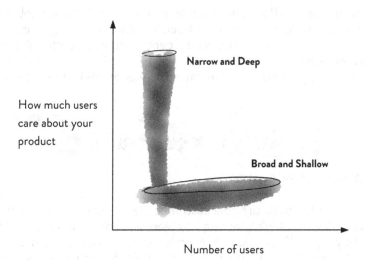

Figure 4-2. Broad and shallow vs. narrow and deep MVPs

Graham's suggestion is to ask the following questions, to measure an idea: Who wants this right now? Who wants this so much that they'll use it even when it's a crappy version made by a two-person startup they've never heard of? If the answers cannot be found very easily, then the idea is probably not going to work.

The point of an MVP (minimum viable product) is that the product idea can be validated. The MVP must be both minimum and viable, meaning that it should represent the hypothesis as much as possible while keeping the effort invested as small as possible. Simply building a small product that is irrelevant is not going to provide much useful information. At the same time, viability of a product doesn't appear magically. Otherwise, every product would be a smashing success. The key is that there needs to be a balance between minimum and viable.

So, how do you build an MVP? The answer is to find a specific (narrow) job to be done and create a solution that addresses that specific job really well (deep). The key here is to go deep and stay narrow, meaning that the solution should help people achieve the goals they hire the product to do in a significant way, but at the same time, limit that group of people to a very specific segment or niche. For example, Groupon started as a Wordpress blog geared to a group in the Chicago area buying coupons. The targeted segment was limited to a specific geographic area. In the case of Spotify, the MVP was a prototype music player that contained a few songs but played music instantly and smoothly, to be shared among family and friends. In Slack's case, the company first focused on serving engineers at technology companies—a niche audience compared to the users and industries it serves today.

The key is that the MVP should be a real product that fulfills the job that the customer needs done at a minimum level. Initially, the main goal is to learn. With each round of feedback, the product gets better and better. Author and consultant on agile and lean methodologies Henrik Kniberg came up with a chart, shown in Figure 4-3, that illustrates the way an MVP should be built.[4]

Figure 4-3. How an MVP should be built

Shipping one wheel and a tire to a customer, then a chassis, then a windshield is a process that wouldn't generate any feedback and assumption validation. This is not how MVP works. In Kniberg's example, the car does the job of getting the customer from point A to point B. Shipping a skateboard first allows the customer to get from A to B, giving the customer the opportunity to try the product and then relate feedback on how the skateboard does or doesn't do the job of getting from A to B. Through this process, we can discover that the car might be the ultimate solution for the job. Or we could discover that a bicycle does the best job, or that even a car might not suffice,

[4]CRM Team, "Making Sense of MVP (Minimum Viable Product)," YouTube video, www.youtube.com/watch?v=0P7nCmln7PM, 11:46, August 1, 2016.

a plane being what the customer requires. The point is that only through this process can we gain the insights and feedback that will inform the final product that satisfies the customer's goals and provides a great experience.

Painkillers vs. Vitamins vs. Candy

A similar approach to evaluate product ideas is to differentiate them by using the painkillers, vitamins, and candy analogy coined by venture capitalist Kevin Fong.[5] According to Fong, startups should always throw away candies, because they only look at vitamins, and really focus on painkillers.

This ethos only makes sense if it's not taken literally. Some people's lives don't depend on candies. However, to those with an especially strong sweet tooth, candies can be viewed as painkillers, as their quality of life will significantly decrease, if candies are denied them. The point is to figure out what kind of painkiller your product is and for whom that is the case.

If you can't determine the answers to these two questions, the idea probably isn't good enough.

Build, Measure, Learn, and Iterate

After the initial idea has been selected, the team focuses on making that idea a reality. The goal at this point is not to sell the idea and make a profit right away, because chances are that the product is not serving customer needs yet. More likely than not, the kind of benefit imagined doesn't necessarily fulfill customers' actual needs. Therefore, there is a high probability that designers need to start over, because the idea is so far off target. Embrace that uncertainty. Because through iteration, we can get closer and closer to creating a good product.

Build-Measure-Learn is an important process in the Lean startup product-development methodology developed by Eric Ries,[6] a startup advisor, author, and entrepreneur. Ries uses this method to help startups, by combining hypothesis-driven experimentation's iterative product changes with validated learning from real users and customers.

The core benefit of using the Lean method is that it shortens the time to produce an MVP and that to a product/market fit—an elusive place in the market, in which the startup's product has proven to satisfy the customer's core need, and the startup can focus on scaling and growing the product.

---

[5]Stephen Fleming, "How to Get Startup Ideas," Academic VC, http://academicvc. com/2007/06/04/painkillers/, June 4, 2007.
[6]Ash Maurya, *Running Lean: Iterate from Plan A to a Plan That Works* (Sebastopol, CA: O'Reilly Media, Inc., 2012).

Along with the Agile software development method, the Lean process has been adopted by almost all Silicon Valley startups. It has become the de facto way to build products, as they exit the idea generation phase into the idea evaluation phase. The core concept to the Lean method is the iterative process of build, measure, and learn.

In measuring customers' and users' reactions and behaviors against the MVP, the team makes adjustments or even pivots altogether. This process is illustrated in Figure 4-4.

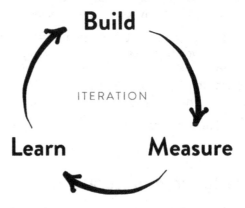

Figure 4-4. Build, measure, learn

A Note on Welcoming Environments

While it is important to be analytical and critical in the build, measure, and learn process and move as quickly as possible, idea generation should be inclusive and welcoming, because the best idea can come from anywhere. This is not to say that a product should be built and scoped by a committee. It is that the product team environment should be welcoming, to the point that everyone feels empowered to provide their thoughts and hypotheses.

Product designers should be responsible for helping to create an environment in which ideas are welcomed and cynicism is never allowed. The yardstick measurement is that no one should feel worse for having offered an idea.

Also, enduring, world-changing products have and can never be built from the minds and efforts of lone geniuses. When great teams are in place, the whole is always much greater than the sum of its individual members. Often, teams can be very large. Therefore, it is crucial that designers lead the team in creating an environment in which good ideas can surface from anywhere and be contributed by anyone. Because product designers hold that special key to the door that connects the product definition phase to the user experience definition, they especially should strive to help the team create a nurturing environment.

Jonathan Ive provides a good explanation of why ideas must be nurtured.[7] This is because every significant innovation seems like an impossibility at the outset.

> *Steve used to say to me—and he used to say this a lot—"Hey Jony, here's a dopey idea."*
>
> *And sometimes they were. Really dopey. Sometimes they were truly dreadful. But sometimes they took the air from the room and they left us both completely silent. Bold, crazy, magnificent ideas. Or quiet simple ones, which in their subtlety, their detail, they were utterly profound.*
>
> *And just as Steve loved ideas, and loved making stuff, he treated the process of creativity with a rare and a wonderful reverence. You see, I think he better than anyone understood that while ideas ultimately can be so powerful, they begin as fragile, barely formed thoughts, so easily missed, so easily compromised, so easily just squished.*

In product design, there is no room for cynicism. Let all ideas develop and have their chance to be validated. You never know which ones are the skateboards that eventually get turned into cars.

Research and Data

In Chapter 2, the importance of having empathy was introduced. Empathy is a tool that product designers should deploy, in order to embody the thoughts and emotions of the people for whom they are designing. Empathy should be an imbued trait of a product designer.

However, there are two more skills related to understanding users that a product should master and deploy during the measuring and learning phases of creating a product. The first is the ability to plan and conduct user research and analyze the results. The second is the ability to work with data scientists or data engineers and the product manager, to understand and interpret data.

User Research

User research is the work related to understanding users. It is the work involved in understanding the goals, motivation, behaviors, and modes of thinking of actual and prospective users.

---

[7]Jonathan Ive, "Jonathan Ive—Tribute to Steve Jobs," YouTube video, www.youtube.com/watch?v=GnGI76__sSA, Threecubed, 7:29, October 24, 2011.

User experience boils down to the touch points that a user has with a product, a company, or a representative entity of a product or company (an e-mail, a product package, etc.). For startups, there are three main ways in which user research can impact the product.

First, user research can be used to validate the merits of an idea. In this case, user research can be conducted to address the question, Is this idea worth pursuing? Second, research can be used to evaluate a tangible realistic (-looking) product, to see if people gravitate toward certain features. The third type of research focuses on a long-term examination of how people interact with a product, to measure whether it met the identified need and is usable. Table 4-1 provides more details on how user research can best serve a startup.

Table 4-1. Three Types of Research

Stage of a Product	MVP	Post-MVP	Long-term
	Validate a concept	**Evaluate** performance	**Measure** change over time
Questions to be answered with research	*What is it?* *Is it worth pursuing?*	*Are people using it?* *What are the areas for improvements?* *Can we prove our hypothesis?*	*What is the impact of this product?*

The first step in conducting a user research session is to identify what your target audience considers a unique characteristic. Do they shop online? Do they have certain interests? If your product is meant for everyone, to the point that no unique characteristics can be gathered from prospective users, your product is not broad enough. Its focus must be narrowed and deepened, as Paul Graham recommends, as cited earlier in this chapter. If you do have a narrow and deep target market, use the characteristics defined to screen for target users.

In brief, user testing can be as simple as giving users a task and watching them complete it. If different users find the design difficult to use, we know that it must be fixed. The key is to get people to talk about why the product was difficult to use and why they made the decisions they did. You can even conduct a usability study on competitor products, to understand what makes them easier or harder to use, so that you can learn from the results.

Data Science

Product analytics provides insights into how successful your product is over time. Tools such as Google Analytics, Mixpanel, or Intercom are all great at measuring user actions and providing analytics about your product. With these tools set up, or if your engineering team decides to roll their own analytics, your team will be able to see what people are clicking or tapping and where they are dropping off.

However, what product analytics don't tell you is the why behind those tracked behaviors. Here's where usability testing can be conducted to understand the whys. Bringing in users and simulating their experiences gives the researcher an opportunity to dig into the why.

It is also useful to track a large set of data points, to enable the data team to analyze and uncover patterns. Product designers should focus on a single metric that really matters at each stage of the product cycle. It's always good to rely on the data scientist's expertise to validate product metrics. Work with the product manager and the data scientist to define that one reliable metric.

Gaining insight from that metric will help to frame the design process with a measurable actionable goal. This doesn't mean that usability and user experience can be overlooked. It means that the design can be created from the get-go with the product's success in mind.

Summary

Ideas big or small can be turned into reality, as long as they have merit and a team has been set up for them to be successful. Product designers are responsible for helping to create a welcoming environment, in which ideas can come from anyone, as well as help product managers keep a product's audience narrow and deep from the start. As the idea materializes, the product designer should conduct research to understand ongoing user and customer feedback at the MVP, post-MVP, and long-term stages. The product designer must also be involved in the definition and measurement of the one metric that will define the product's success at each stage.

Design Is a Team Sport

How to Work on a Startup Team

If you look at the definition of the word *design* as a noun, it will likely read something like this:

> A design is a drawing which someone produces to show how they would like something to be built or made.[1]

While this definition is perfectly applicable to most types of design work, this chapter will expand on this definition, to establish a new definition of design— one that relates to building products in startups in a team setting.

Twenty years ago, *product design* was a term reserved for the design of physical products, and it was often used interchangeably with *industrial design*, which is a term that refers to the design of form and usability to mass-produced products.

When the Internet first took off in the late 1990s and early 2000s, it kick-started a new field of design, one that was concerned with the look and feel of web sites. The Web was a new technological medium and product paradigm

---

[1] *Collins English Dictionary, s.v. "design."*

© Tony Jing 2018

T. Jing, *Hacking Product Design*, https://doi.org/10.1007/978-1-4842-3985-8_5

that quickly gained prominence in the world. As the popularity of the Web grew, a new field of design emerged that focused largely on its usability and user experience. In the late 2000s to early 2010s, as the Internet and smartphones became ubiquitous, user experience (UX) designers became highly sought after.

However, as businesses and the field of design evolved together, it became more apparent that design is not just about ensuring good usability or creating web experiences. It also solves business problems related to the users' goals and experiences and how they fit into a business context. Hence, the focus on business, in addition to user experience, led to the term *product design*.

While many bigger businesses still divide design work into UX, user interface (UI), visual design, user research, prototyping, etc., more and more companies are merging the role of the product designer with others. Especially in a startup, a designer must wear several hats. By their nature, startups are interdisciplinary and fast-moving entities. As introduced in the previous chapters, the nucleus of a startup consists of the product manager (PM), the designers, the engineers, product-marketing managers (PMM), user researchers, data scientists, and, sometimes, project/program managers, customer relationship experts, and legal and operational experts.

Startup designers must be able to work with these people, by capturing their input, feedback, insights, and knowledge, while using design to facilitate and bring people together. A good analogy to describe the startup product-building process could be a relay race in which a team competes with none other than themselves to cross the finish line. The baton is passed back and forth among members of the team frequently, so teammates must run closely together (Figure 5-1).

Figure 5-1. Teamwork in a startup is like running a relay race

The success of the work relationship fostered by the product designer closely correlates with the success of the product. Without a great working model, and trust within the team, it is very difficult to achieve product success. This chapter is an introduction to how product designers can facilitate a working model of an idea type within a startup.

Product Process

One typical process of how a startup team builds a new feature, from a product designer's perspective, follows:

1. Members within the team identify the market and the customer needs, goals, and jobs.

2. The PM creates and distributes a problem statement, based on these insights and data.

3. Feedback is given by the cross-functional team on the problem statement document, and alignment on the problem is achieved.

4. The PM creates an MVP (minimum viable product) feature narrative doc for designers and engineers to know exactly what should be built.

5. Designers create flows, wireframes, and mock-ups based on that feature narrative doc, to propose a design solution through an iterative approach.

6. The stakeholders and the team review the designs and the prototypes and provide feedback.

7. The designs are iterated upon in a few rounds, usually validated by real users' feedback in research sessions.

8. The team agrees on the right solution, and engineers begin working to implement it.

9. Data scientists begin to set up ways to track the approved product, and marketers and the customer relationship managers prepare a rollout plan or a marketing plan.

10. The engineering team finishes building the MVP to be tested in the real market.

11. The product is tested with a subset of the overall customer or user base over a period of days, weeks, or months.

12. If the product proves to reach the goals and the objects, based on the data collected in the test period, the product is rolled out to all existing or new customers.

By this process, a typical product designer is mostly involved in steps 2 to 7. The level of a designer's effort rises between steps 2 and 4, peaks at step 5, sustains at steps 6 and 7, and then gradually dissipates at step 8.

This is only one way in which products are built. However, despite how the process may differ, one thing is certain: collaboration is at the heart of how modern digital products are built inside startups. Once a team is formed, product managers, designers, engineers, and all the other functionaries must learn to work well together.

How to Work with PMs

The most important relationship and partnership that a designer has in a product team is the one with the product manager. A hint of this can be seen in the fact that both titles are about the "product." The product manager defines what the product stems from, as in discovering and defining which problems should be solved and what jobs customers need to have done. Product designers, on the other hand, translate those discoveries into mock-ups and other representations of the product.

Understand a PM's Job

Vice president of product design at Facebook Julie Zhuo[2] put it best: a "PM's job is to be a connector that helps teams ship successful products." PMs are the ones who are supposed to unite the team in the journey of uncovering addressable problems and jobs to be done. The key here is *addressable*. There are endless problems facing the world. However, given the insights and skill-sets available at a startup, that number becomes only a few, sometimes only one. It is the product manager's job to identify what's possible; therefore, it is important that PMs

1. Understand the startup mission, the business needs, and objectives

2. Identify the target user and the jobs to be done

3. Understand the constraints (engineering, organizational, etc.)

4. Communicate points 1, 2, and 3 with the team

[2]Julie Zhuo, "How to Work with PMs," Medium.com, https://medium.com/the-year-of-the-looking-glass/how-to-work-with-pms-3e852d5eccf5, August 2013.

When products fail to deliver, PMs usually take the brunt of the blame. This is because the PM's job is to represent the problems, the goals, as well as the roadmap and the execution of the design and its implementation. In the early stages of a startup, there usually exists only one product, with one PM, who could also be a founder or an early employee. The role requires a clear ability to prioritize and communicate, both verbally and in writing.

The other trait of a successful PM is the ability to work with different people. Depending on the product, industry, and context, a product manager may have to put him- or herself in many unfamiliar situations, interfacing with various functions, within or outside the team and the company. Product managers typically have a lot of convincing to do. They usually must earn the team's trust to get things done, because they usually don't have the power or the logistical bandwidth to hire their own team.

Understand Your Job

A key to establishing a good working relationship with a PM is to really understand your own job as a product designer. While it's the PM's job to articulate the goals of a product, the product designer is tasked to figure out how to achieve that goal.

Inexperienced PMs and product designers often confuse the two. Sure, design, UX, and UI feedback can come from anywhere, but it is just that—feedback. Product designers should not treat them as directives. Just as the PM should be expert at gathering constraints and contexts and then refining requirements and goals, the designer should be expert at absorbing those requirements, as well as suggestions, and evaluate, iterate, and create interfaces, experiences, and systems.

Also, product designers should realize that it is not their role or within their expertise to discover problem areas, jobs to be done, market segments, industry contexts, etc. While product designers should participate in those processes and should feel free to share insights, suggestions, and observations on those subjects, the responsibility of defining the scope of a problem lies on the shoulders of the product manager.

Setting expectations and aligning them with the product managers' early in the working relationship can help to avoid later friction and difficulties.

In addition to the preceding, there are a few other ways in which designers can help PMs do their job. First, product designers should always embody the user's point of view. This is not to say that users are always right or are reluctant to try new things, but, rather, that product designers can be more intuitively disposed to be empathetic to users. That disposition can be harnessed to steer the design focus to the user's goals and experiences, if a PM veers toward solutions that reward the business, or the engineering process, at the expense of user experience.

Second, a product designer must learn to expand the focus beyond the user and shift the conversation from how something is achieved to why there is a desire to achieve something. Ask for the product's desired impact, in terms of business goals and objects. Ask the question, Assuming our end product/ feature does the job that users hire it to do, what business objectives should it achieve, in both the short and long term?

The third way to help the PM is for the product designer to understand the trade-off between prioritization and design perfection. In a perfect world, there would be infinite time to ensure that each problem is solved perfectly. However, we don't live in a perfect world. Therefore, on starting a project, try to figure out the constraints and design baselines as soon as possible, such as what the baseline user experiences should be. This will help to shape the design process in a way that makes feature prioritization a smooth exercise, rather than a painful one, when the need to cut certain features arises.

One of the keys to prioritizing design features with your PM is an impact and ease of implementation[3] table (Table 5-1).

Table 5-1. Impact and Implementation

Impact		
	(4) High Impact Difficult to Build	**(1)** High Impact Easy to Build
	(3) Low Impact Difficult to Build	**(2)** Low Impact Easy to Build
	Ease of Implementation	

It's up to PMs and engineers to determine how a product's features fit into each quadrant and what the necessary trade-offs must be. While the imperatives cited in quadrant number 1 (high impact and easy to build features) should be prioritized first, those in quadrant 2 and 4 should be prioritized in the context of each project.

---

[3]Adhithya Kumar, "A Designer's Guide to Working with Product Managers," UX Collective, https://uxdesign.cc/a-designers-guide-to-working-with-product-managers-bcb164a473df, February 23, 2017.

Handling Disagreements with Your PM

As mentioned earlier, while the MVP is what the team strives for, don't allow that be an excuse to let the core usability and utility of your product suffer. Know what is really important and what the team is trying to prove or disprove about the feature you're about to ship.

The best way to do this is to treat your PM like a partner from the get-go. Building a channel for open communication is key to getting a faster resolution to an issue. When it comes to decisions related to user experience, your PM should defer to you. What the PM should do is articulate the importance of the product's goals and the desired time line from a business and customer-focused perspective. The PM should present convincing evidence, rather than giving orders on what the product should be or should look like. From this perspective, PMs are first and foremost problem managers.

If you can't agree on how to work together, you should seriously consider switching to a PM who respects your model. Aside from this, there is one additional factor to which designers should pay extra attention.

This involves when a product is deemed ready to be shipped. The key here is to balance what the best experience to be delivered to the user is, to ensure that business goals are met—whether those goals are to move a metric or prove a hypothesis. Understanding the goal is crucial for the designer, and that is something design should relentlessly define and refine with the PM. With that goal set, the designer should develop a framework of how the design performs against attributes that are important to achieving the goal. Initially, the measurements can take the form of a vote among project stakeholders who have context and, later, the form of user testing with actual users and customers.

How to Work with Engineers

Without engineers, products don't get built. Also, unlike PMs, designers must support engineers by (a) explaining the design, to ensure that engineers understand what it is they are building, and (b) creating digital assets that engineers will require to implement the designs.

Understand Their Job

In technology startups, engineers are the ones truly translating a vision or sketch into something real. This should not be overlooked. It doesn't matter how good your design is, if you don't have buy-in from the engineers who are building it, that is, making it real, reliable, fast and slick, presentable in all devices and viewports, localized in different languages and cultures, and scalable to millions and even billions of users.

Just as your PM shouldn't treat you like a "resource," you shouldn't treat your engineers like one either. Instead, treat them like partners, and help them to understand the most critical things you observe.

Know Some Programming

Start by knowing the basics of how programming works. The goal is not to be able to code as well as engineers but to speak their language and build empathy for the work they do.

Engineers spend their time figuring out how to not repeat themselves in terms of writing the same logic and how make their code scalable to the product. Designers should think in systems and patterns and have a wealth of design resources to explain their work, so that engineers will eventually realize their designs.

For example, designers don't have to know how to write HTML, CSS, and JavaScript, but knowing what the relationships between these languages are can help designers understand and recognize the limitations of what's possible on the Web. This saves time and effort for both designers and engineers, because both parties can operate within the same field of reality, on the same page, to determine what's mundane, innovative, or too far-fetched.

Get to Know Their Constraints

Aside from knowing the baseline technology constraints, it's advisable to bring engineers into the design process as quickly as possible, as long as the feedback you're looking for is framed correctly. When it comes to specific patterns or interactions that have not been implemented before, it's always good to check in with engineers, to see how long it might take to realize them, or if they're even possible.

By involving the team in the design process, feedback can be provided in stages, to reveal technical issues and constraints. Early in the process, having alignment on the flow of the product could help to unblock back-end engineers, as they architect how the product could work behind the scenes. Getting the conversation rolling allows engineers to work out solutions to technical challenges among themselves and postulate possible workarounds.

Always be proactive and upfront about what is changing and what is locked for the current sprint. Engineering work is a lot harder to redo than design, owing to the various factors being considered—scalability, stability, localization, etc. No engineer will expect that all of her/his work will be shipped in the final product, but we should keep the amount of throwaway code to a minimum. Taking the example of construction, it's worst when you've already laid down the foundation and the walls, only to realize that the designs for the walls must be altered to support a change, or that the foundation has to be redone as well. Perhaps it wouldn't be so bad if the deadline and/or budget could be adjusted to accommodate these changes, but in the real world, they usually can't. So, work with the engineering team and always do as many iterations as possible before locking in a design. Once something is locked, commit to it, by holding your PMs accountable.

Again, Guide Them to the Users

Everyone who works on products loves to see their work truly benefit end users. Engineers are no exception. When engineers see the work they do in action, they can then relate to the whys and hows of a design. They can see the rationale and value of their work in the world, rather than simply blindly executing it.

While product managers are often wrapped up in navigating feature priority, time line, and metrics, designers have the unique opportunity to relay the voice of the user or customer.

It's good to make user-testing feedback bite-sized and easily digestible, especially in the form of video or audio. This way, anyone can quickly grasp the ins and outs of the product and see where it shines and where it falls flat. Sometimes, to get products to really shine, user feedback lights the fire of motivation.

Design's Superpower

Designs have the unique ability to make something look and feel real in a quick amount of time. In any organization, that ability, when applied correctly, can connect the teams and stakeholders and inspire people to work together with a shared context and insight.

Everyone Will Have an Opinion

Ideas can come from anywhere, and they often do. However, most ideas aren't developed to their full form, or even beyond first inklings. Therefore, ideas are fragile.

One of the best things designers can do is work with people on their ideas, from everywhere within an organization. One of the magical powers of designers is the ability to "fake" things—sketch things out, put together something quickly, or build prototypes. This is not unlike how an origami master can create the likeness of a swan, based on a real swan, by folding paper, as shown in Figure 5-2. Designers should have the ability to turn ideas into prototypes that bring those ideas to life. Doing so will elevate relevant conversation to a more tangible level and help to uncover differences and gaps in each person's own interpretation of the supposedly same idea.

Figure 5-2. An origami swan could be regarded as a prototype

Prototyping ideas will help the design team to increase buy-in, encourage interesting perspectives, and allow for an artifact to promote real discussions with realistic expectations about where progress and improvements can be made. That is one example of a designer's superpower to make realistic contributions and create positive change in any organization.

Summary

Beyond simply creating flows, diagrams, UIs, and visuals, a large portion of a product designer's job is to collaborate with product managers and engineers, so that they also get to understand the user's goals and perspectives. Product designers have the unique responsibility to help both the problem definition and implementation stages of the product-development process, to ensure that the product being built meets the jobs users are looking to get done, in a usable and viable way. Designers should always realize that they have the special power to make ideas come alive by prototyping. The result generates excitement about new ideas and shifts the conversation from abstract imaginations to evaluating tangible solutions.

Design Is About Priorities

Especially for Startups

One of the core differences between a product designer in a startup and a product designer in an established company is the spectrum of responsibilities. Startups are the very definition of chaos. This is the case even in the most organized and well-run startups. By their nature, startups are the manifestation of a group of people not knowing whether something will succeed but trying it anyway. Some of those people provide the money (investors); others provide the skills (founders and employees).

There Is No "Right Time"

For most startups, product roadmaps and strategies are constantly being worked out. No one has a recipe for success, and as a result, no one can be sure of the "right time" to do most anything.

But that's okay.

As product designers, we should realize that we're all in this together, trying to figure this thing called product out. So, be ready to remind people to step back to think, not only of the business, but also of the users, their contexts, and the society as a whole, as Chapter 5 highlights. Work with the founders

© Tony Jing 2018
T. Jing, *Hacking Product Design*, https://doi.org/10.1007/978-1-4842-3985-8_6

and key decision makers to embrace empathy for the users, as they define the value proposition of the product.

The truth is that the right time is often now. Action begets action. Once an idea is agreed upon, go ahead and mock it up. Present your mock-up to five people and observe how they react. This is the beauty of working in a startup: you can always find ways to help shape and define the product and the strategy. Oftentimes, you don't even need a plan. Simply start acting, and your actions will illuminate the logical steps to follow. The key objective of taking action in a startup is to validate ideas as soon as possible. This is something that must be understood and internalized by product designers.

Reid Hoffman, the founder of LinkedIn, said that building a startup is like jumping off a cliff and assembling an airplane on the way down.[1] This is not a far-fetched statement. Startups are given a limited amount of money to prove an idea and to use that idea to make a business work. Each month, a (sometimes large) portion of that money is spent to keep the startup going forward—paying its employees, its suppliers, and for overhead. It's often the case that months will go by without profits coming in, as the company and team have yet to crack the predicted market.

This is why time is of the essence. In startups, the designer is often the researcher, and the research work has to be done throughout the design process. They key is that designers should realize that fast and good don't have to be mutually exclusive goals, as long as they realize that good doesn't have to be perfect. It simply means good enough is fine, as highlighted in the sample MVP discussed in Chapter 4—building a skateboard first, then a scooter, and then a bike, etc.

So, be ready to make trade-offs. Getting to good enough is all about making the right trade-offs to achieve the maximum.

If the product is still in the MVP stage of its life cycle, meaning that the main feature hasn't been borne out by customers, priority should be placed on the main feature. That is, the nice-to-have feature(s) should be tabled until later.

More concretely, this means that every design decision should be subjected to validation, which could be quick user-testing sessions with users who match the target audience or more formalized user research sessions with candidates from the product's actual user base. Don't be afraid to reach out and simply ask your first users to give your product a try or ask them what they think about it. It is often less time-consuming than we presume, and the results are almost always very insightful.

---

[1]Reid Hoffman, "How to Be a Great Founder with Reid Hoffman (How to Start a Startup 2014: Lecture 13)," Y Combinator, www.youtube.com/watch?v=pkAum45ubWc, YouTube video, 49:57, filmed 2014, posted April 2017.

One key difference between startups and big, established companies is the pace of change. Being able to change, adapt, and evolve quickly is the hallmark of all great startups—and product designers. This ability to change quickly is about being comfortable with uncertainty about a product's roadmap and the outcome of your designs. Pivoting is a real thing. Rather than fighting it, embrace the change, be agile, and keep learning beyond what is required.

The Only Metric That Matters

When it comes to each new feature, each new flow that needs to be built, the seasoned PM-turned-VC Josh Elman argues that there is only one key question that product people should ask, which is, "How many people are really using your product?"[2]

Elman points out that most of the fluffy numbers, such as page views and monthly active users, don't reveal much. Instead, the only question that matters is, Who is really using the product? The answer to this question should be defined in the form of "x amount of users did these things within this time frame." An example of this for a hypothetical search engine product can be "100 users performed three or more searches within the last 15 days."

Elman is speaking from a social product perspective, with a focus on growth. However, the core of his assertion is true for all products: a product should have a clearly measurable and contextual objective that contributes to its long-term goal. Twitter's mission is to "give everyone the power to create and share ideas and information instantly, without barriers." Through examining how people used Twitter over a period of time, Elman found that in order to give people that power, as mandated in the company mission, a new user has to have visited Twitter at least seven times in a month, then it becomes considerably more likely that user will stick to Twitter.[3]

The product team working with this problem extrapolated a new goal of aiming for this metric of seven times a month and devised ways to provide enough value to new users, so that they would come back to the app at least seven times monthly.

Frankly speaking, there are many metrics that can indicate the health and context of a feature in acute yet precise ways. For a larger company, the number of metrics may vary greatly, and each may carry significant importance.

---

[2]Josh Elman, "How to Work with PMs," Greylock Partners, https://news.greylock.com/the-only-metric-that-matters-now-with-fancy-slides-232474cf414c, June 2015.
[3]Ibid.

In a startup, however, it's a different story. The startup product team has to deliberately choose to not focus on all of the other metrics, because the opportunity cost of not focusing on the key metric, the one that really matters, is simply too high. A startup doesn't have the time and resources that established companies have to address a problem from a multitude of angles.

The metric doesn't always have to be as detailed as shown in Elman's examples, but the key is to give context to a number. Are people really using your product? That's the question we need to address. The reason is that the product team can use it to figure how to really help, add value, and benefit the people buying and using the product.

Defining it with the product manager can inform the short-term and long-term design goals. Knowing it will also help to clarify lesser goals and frame design decisions. At the very least, having this conversation with the PM can help to uncover and evaluate preexisting assumptions.

Apply the 80/20 Rule

The 80/20 rule is that often, in a large system, 80% of the output is caused by 20% of the input (Figure 6-1). You see this pattern repeatedly, whether in economics, transportation, human habits, etc. Here are some examples:

- 80% of traffic is on 20% of the roads
- 80% of a company's revenue is generated from 20% of customers
- 80% of the product experience comes from 20% of the UI

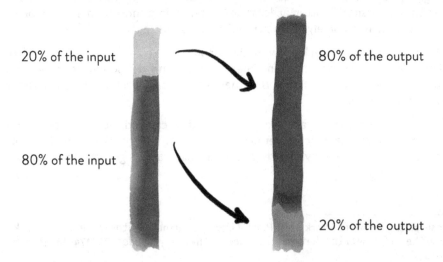

20% of the input

80% of the output

80% of the input

20% of the output

Figure 6-1. The 80/20 rule

This pattern is also known as the "Pareto principle," which is a rule of thumb for denoting power law distribution. The takeaway is that having this rule in mind allows us to define the order of importance of the problem at hand. It allows us to focus on finding and pulling the right levers that move us toward the goal, one at a time, deploying our limited time and resources to uncover and address the most important areas at each stage.

One story that comes to my mind is Tian Ji's horse race. This is a Chinese idiom about a general during the Warring States period (475–221 BC) who won a best-of-three horse race despite having slower horses.

Essentially, Tian Ji gathered intelligence on his opponent's horses before the race and deployed his horses on race day in a way that matched his best horse with his opponent's medium horse, and his medium horse with his opponent's weakest horse (Figure 6-2). The result was that Tian Ji narrowly edged out his opponent in both matches, despite losing the race between his weakest horse and the opponent's strongest horses. He won the best-of-three series two races to one.

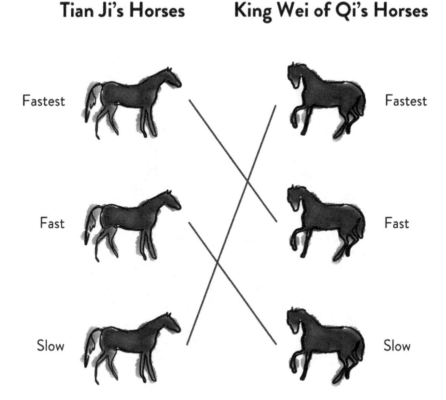

Figure 6-2. Tian Ji's strategy

The lesson here in regard to the 80/20 rule is to (a) get to know the situation, the problem, resources, and constraints; (b) identify the 20% area you should be spending energy on and deploy your resources to it; and (c) repeat the process of gathering intelligence, strategizing for the next stage. This process of solving one critical issue at a time can extend from business strategy all the way down to individual UI decisions.

Summary

Given the nature of startups, product designers must be decisive and have a bias toward taking action. In addition, designers should work with product managers, engineers, and data scientists to uncover the one metric that truly matters when growing a product. Tactically, designers should understand and use the 80/20 rule, to prioritize and make decisions.

Designing for Scale

From Four Perspectives

The word *scale* has been used in the tech world to mean many different things. In engineering, it refers to the process and methods that can lead the product or service to handle often-sudden increases in system complexity, user requests, and actions. It enables the product to continue to perform smoothly, without defects and failures as the company "scales" bigger.

Scale has become a verb that founders and investors throw around in discussing the growth of a startup. In general, it means the conscious effort to grow something. Whether that something is the number of users, the revenue numbers, the profit margin, or the regions covered by a product is entirely dependent on the context of the company.

In this sense, to scale a product means to adapt sudden growth—in usage, number of markets and languages localized and supported, complexity of the product's internal workings, and depth and width of features, new and old.

While this take of the word *scale* is something that startup designers should understand and work with, there are three more layers of the word that product designers should embrace.

© Tony Jing 2018
T. Jing, *Hacking Product Design*, https://doi.org/10.1007/978-1-4842-3985-8_7

1. Design is about trade-offs, but that doesn't mean it should be limiting or elitist. Good design should be egalitarian; it should benefit as many people as possible.

2. To scale a product is to not confuse product aesthetics or product user interface with the product itself.

3. Always design with the future in mind. Rather than creating feature after feature, create systems and processes that can help everyone adapt to changes. Use design to empower within.

Fall in Love with the Problem

Growth creates problems.

Many insects go through ecdysis—the periodical shedding of old exoskeletons as the insect grows. Human beings experience a period of fast growth during adolescence, yet that time is accompanied by awkwardness.

Startups are no different. Sometimes the only way to go forward is to shed old protocols, ways of doing things, and even technology stacks. Sometimes scaling means handing a variety of problems that are the result of growth.

The designer's job, with input gleaned from user researchers and data scientists, is to help product managers, engineers, and marketers make sense of the increasing complexity a growing product faces, through the lens of human experiences and product potentials.

To do so is to ask questions. For example, As our product grows, how is our customer's experience being affected? What are the positive and negative extremes of that experience? How are the experiences of individual customers and users distributed? Do they all skew toward the positive extreme, the negative extreme, or the middle? These are the questions that designers must ask. These questions can help the team realize the user experience risks associated with increased complexity and eliminate user experience blind spots.

In addition to these questions, the designer should lead the product team in painting a picture for the future—a vision for what the product can be—by collaborating with product and engineering leads, product leads to provide insights on the company's overall strategy and roadmap, and engineering leads to illuminate possibilities on the technology horizon. Having these conversations and achieving alignments with these leads will help designers in their creation of artifacts (for example, mock-ups and prototypes) that showcase the product's potential as it scales up.

Presenting these artifacts can help to rally the startup team, to ensure that people are not just blindly scaling and adding complexity but doing so with the same goal in mind. As alluded to in Chapter 6, the ability to quickly create mock-ups and prototypes is the superpower that designers have to provide clarity and leadership.

There's an old saying that good designers fall in love with the solution, but great ones fall in love with the problem. This is the one principle that designers can embody to show the team, by example, what the ultimate goal is. Growth is great, but it should be a derivative and a by-product of your product creating value, through solving people's problems.

A startup designer's most important role in a period of growth and scale is to remind colleagues not to lose sight of what is actually important—solving problems for people. In fact, the more complex the problem, the more dedication and care the designer should bring to the table. Teammates will sense that passion and perseverance. Many of them will embrace it, too, and not deviate from the goal of growth—which is to scale the positive impact that your product brings.

Benefit As Many People As Possible

People are different. Their memories are different, their personalities are different, and their perspectives are different. However, in order to create products that scale, designers should focus on how, in the end, people are similar. Our job as designers, when it comes to scaling products, is to focus on the ways people are similar, realizing that human nature gives us similar drives and wants. In Chapter 8, I'll go through some of the universal design principles related to human nature that designers should understand in order to build intuitive products that can be easily used.

Designing for Accessibility

It has been said that one in every seven people in the world lives with some sort of disability. Disabilities can arise from birth, aging, accidents, or health conditions. There are also those with temporary disability, for example, limited mobility caused by pregnancy, or limited hand usage owing to a broken wrist. There are also longer-term disabilities, such as vision impairments, hearing impairments, color-blindness, and impairment of the spine and limbs.

This is a topic that is closely intertwined with that of scale. Products should help and benefit people and be as far-reaching as possible. To accomplish this, designers must incorporate accessibility as an integral part of the user experience.

According to the Centers for Disease Control and Prevention[1] (CDC) in the United States, any person who experiences difficulty in the following activities may be considered as having a disability:

1. Vision

2. Movement

3. Thinking

4. Remembering

5. Learning

6. Communicating

7. Hearing

8. Mental health

9. Social relationships

When examined closely, even the words *disability* and *impairment* imply abnormality and the need for those aspects of an individual to be fixed. While this perspective may help medical practitioners in curing disease, alleviating pain, and enhancing physical functions for their patients, our society shouldn't necessarily adopt this perspective. The truth is that our human environments have not been created equally for everyone.

Rather, designers should help society establish a mindset that recognizes the challenges and barriers that people with disabilities face with the environment, people's attitudes, and other social factors. Specifically, the designer should champion the removal of barriers for all people, to improve their quality of life. This is the fundamental goal of creating products with accessibility, an integral part of design from the start.

Two main trends in world demographics are also making it imperative to create accessible and inclusive designs. The first is that the world's developed populations are all living longer lives. With this, there is an increase in age-related disabilities, such as limited sight, hearing, and mobility. Products that have developed in emerging markets are feeding information back to developed markets. The other trend is that as products become more digital and less dependent on the physical world, there are more opportunities for people to connect and participate in activities that are not limited by the physical world.

The Web Content Accessibility Guidelines[2] (WCAG) were created by the Web Accessibility Initiative, which is a project of the World Web Consortium,

---

[1] Centers for Disease Control and Prevention, "Disability and Health," www.cdc.gov/ncbddd/disabilityandhealth/disability.html, August 1, 2017.

[2] W3C Web Accessibility Initiative, "Web Content Accessibility Guidelines (WCAG) Overview," www.w3.org/WAI/standards-guidelines/wcag/, July 2005, updated June 2018.

an organization and community that establishes standards for the constantly evolving Web. Although WCAG was made for the Web, its principles also apply to products. Four main ideas are promoted in the WCAG.

1. Perceivable

2. Operable

3. Understandable

4. Robust

Together, they form the acronym POUR. Ruby Zheng,[3] a writer and UX consultant at the Interaction Design Foundation, cited the examples of TED Talks videos, Apple's iOS AssistiveTouch, Twitter error messages, and responsive web design to describe the POUR principles.

For the "Perceivable" principle, Zheng notes that each TED Talks video includes a transcript, often available in multiple languages. This makes each talk accessible to people with hearing impairments. The availability of multiple languages scales the reach of TED videos even further, as shown in Figure 7-1.

Figure 7-1. Transcripts are available below videos of TED Talks

[3]Ruby Zheng, "Understand the Social Needs for Accessibility in UX Design," The Interaction Design Foundation, www.interaction-design.org/literature/article/understand-the-social-needs-for-accessibility-in-ux-design, January 2018.

For the Operable principle, Zheng explains how Apple gives its users the option to choose alternative methods of navigation that do not require pushing buttons physically, as shown in Figure 7-2.

Figure 7-2. iOS AssistiveTouch

Reflecting the Understandable principle, Twitter's error validation messages for its sign-up page were highlighted by Zheng to illustrate the ease in which information is displayed. Twitter's inline validation messages are displayed in red and right beside the corresponding input fields, thus allowing users to quickly understand and correct issues. Figure 7-3 is an example of how Twitter achieves this.

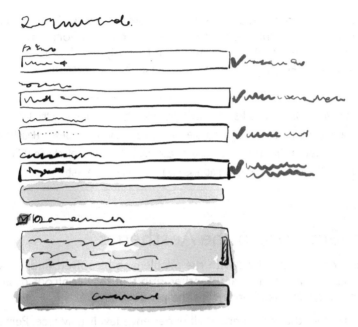

Figure 7-3. Twitter input validation

For the principle of Robust, which characterizes the reliability and adaptability of content to a wide range of devices and screen sizes, Zheng cites as a good start the emergence of responsive web design, that is, the work of designers and developers to create easily accessible web sites, regardless of the device constraints. Figure 7-4 illustrates the concept of responsive web design.

Figure 7-4. Responsive web design

The key takeaway from these examples is that while accessible design can take many forms, it is up to the designer to understand the opportunities each circumstance and context allows for, and then to apply the POUR principles, to produce designs that are as inclusive and accessible as possible.

■ **Note** If you're building a web-based product and want to deep-dive into the techniques and principles of designing for accessibility on the Web, please check out Jesse Hausler's article[4] "7 Things Every Designer Needs to Know about Accessibility" on the Salesforce UX blog. It is a great overview of the basics of designing accessible interfaces for the Web.

Don't Get Lost in the Aesthetics

Aside from not falling in love with the problem to be solved, and not designing for inclusivity from the start, another trap that may confront designers is confusing a product's aesthetics with the product itself.

This is a problem that designers at all experience levels may face. Perhaps the root cause of this has to do with the fact that the majority of interfaces with which humans interact with digital products are visual-based, and most product designers either begin their careers as visual or graphic designers, or they simply have a special affinity for the visual aspect of their job.

Regardless of the reason, time and time again, we see products that look amazing on the surface but actually don't help and benefit people. Paul Adams,[5] VP of product at Intercom, a platform for companies to acquire, engage, and support customers, coined the term *the dribbblisation of design* to describe this phenomenon.

Dribbble is an invitation-only web site on which visual designers can post their work, for discussion with other designers. The posts are usually in 800 by 600 pixels and cover graphic design, illustration, UI design, sketching, and motion design (through short animated GIF files).

Adams discusses the problems he witnessed as a hiring manager for product designers at Facebook and Intercom. The core issue he noticed was that "too many designers are designing to impress their peers than address real business problems." What Adams pointed out was the proliferation of seemingly

---

[4]Jesse Hausler, "7 Things Every Designer Needs to Know about Accessibility," Salesforce UX, https://medium.com/salesforce-ux/7-things-every-designer-needs-to-know-about-accessibility-64f105f0881b, April 15, 2015.

[5]Paul Adams, "The Dribbblisation of Design," Inside Intercom, https://blog.intercom.com/the-dribbblisation-of-design/, May 21, 2018.

beautiful apps and software that ultimately look and feel the same, because they all sample the same styles that don't address the issues facing real people.

How many of the hundreds of weather apps, dashboards, e-commerce sites, and marketing page designs posted on Dribbble are actually usable and useful? The answer is very few.

Most of these designs claiming to incorporate POUR values fall into the trap of confusing the aesthetics of the product with the product itself. These snapshots of products exist outside of constraints and contexts and should not be evaluated as actual products. If this phenomenon is only limited to sites such as Dribbble, it would not be such a big issue. The problem is that designers are using these mock-ups as good examples of work and, worse, they are doing so in their jobs.

Too often, a designer dives into a project wanting to create something that looks and feels beautiful. Beauty is never the goal. The goal for a startup designer is always, first and foremost, making sure the product does what its customers and users expect it to do. That alone is difficult enough to achieve. Never assume that making something beautiful or slick has anything to do with the value of the product.

Instead, ask why people are using the product and what they really care about. Even though Steve Jobs was not a designer, he summed up this point very succinctly[6] when asked what design is.

> Design is a really loaded word…we actually just talk about how things work. Most people think it's how they look, but it's not how they look, it's how they work.

Design is about making products that work right. Product design in startups is especially so. This does not mean that design is not important. However, as long the design does not take away value from what the product sets out to do, good enough sometimes means perfect.

The worst is when designers get too hyped up about the powers of their craft. They fail to understand the true value of their product. Following is an example of the severe consequences that arise from confusing aesthetics with the product.

In the mid-2010s, a fast-growing social media startup launched a new version of a product for countries in the Indian subcontinent. Tens of millions of dollars were spent in creating this version and marketing it to users. It generated considerable fanfare and buzz before launching. However, months in,

---

[6]Steve Jobs, "Apple['s] Steve Jobs on Design," Dirk Beveridge, www.youtube.com/watch?v=sPfJQmpg5zk, YouTube video, 02:05, posted October 16, 2010.

the results were dismal. Users left one-star reviews, and many simply stopped using the product altogether after a few weeks. In the end, the company had to pull the plug on the new version and pull back to the older version of the product. Even then, some of the users did not return.

This was a devastating business loss. It set back the company's progress in its former markets by years. The reason that users didn't like the new update was because the new version was painfully slow. The initial product loading time took much longer than previously, and the app generally performed very sluggishly, sometimes it would even slow down the operating system of the user's device.

How could that have happened? It turns out the designers of the new version of the product wanted to create the slickest app transitions and animation. Indeed, they were successful. Together with engineers fully on board, the team created the slickest and flashiest transitions that would make any designer ooh and aah with excitement. However, two important things were overlooked.

First, because the transitions were very hard to build, the engineers had to rely on the processing powers of the latest mobile hardware. Second, the designs assumed that data could be loaded as soon as the app opened.

In reality, their users in the Indian subcontinent lacked the latest mobile hardware or software. Most had phones that were at least a year old, with operating systems that were a few versions, if not dozens of versions, behind the latest one. Many of the users also did not have 4G or 3G connection, meaning that they could not pre-download the data, as required by the new version.

The result of these two factors caused the product to provide a slow and sometimes completely unresponsive experience—a huge letdown following the fanfare, and a huge step backward from the previous version of the product, a version that despite being old, still worked.

Of course, the engineering and QA teams bear some responsibility for this big mishap. They should have rigorously tested the product on all devices and noted the subpar performance on older devices. However, the bigger fault lies with the designers of the product. They should have studied the market and understood the persona they were designing for.

A persona is a character, created by designers and researchers and based on in-depth research, that represents a person who might use the product in a typical way. A persona does not represent an average of all possible user traits, characteristics, and preferences; rather, it is an example of what a real user of the product could be like. The work to research and create personas generates a lot of empathy and insights into the target market and audience.

This work was either missing or largely overlooked when the new product was designed. Instead of understanding the contexts and environments of real users, the designers focused primarily on aesthetics, thus ultimately leading the user's experiences and the product's movement in that market into oblivion. This was a costly mistake, and similar mistakes are still being made by startups big and small. As designers, it is our job to ensure that these don't happen. We still have a long way to go.

Systems and Processes

Deep in Yunnan Province, at the southern tip of China's mountainous region bordering Myanmar and Laos, stand dozens and dozens of large wind turbines. Each turbine rises 80 meters atop the 2,300-meter to 2,700-meter mountain range. Each blade of the turbine is 52.4 meters in length, with a weight of 80 tons. High winds effortlessly turn the turbine blades, generating a crescendo of undulating electricity into the local power grid, as shown in Figure 7-5. This is a clean and sustainable source of energy, benefiting the surrounding towns and communities, without damaging the lush natural environment.

Figure 7-5. Large wind turbines on a Yunnan mountain range

Thousands of miles away, in an unnamed European city, a small girl gazes at a MacBook Air inside an Apple store. She effortlessly opens the display lid and sees the black screen light up. Hearing her mom calling her over, she steps away, leaving the MacBook Air on the table, as shown in Figure 7-6.

Figure 7-6. A MacBook Air

What do these machines that are completely different in size and functionality have in common? They are both designed outcomes of complex systems and processes.

Here's why. Years ago, I watched a documentary about design called *Objectified.*[7] In it, a quote by Jonathan Ive, Apple's lead designer, stuck with me.

> So much of the effort behind a product like the MacBook Air was experimenting with different processes. There's a completely non-obvious way you get from this part to this part. There are incredibly complex series of fixtures to hold this part in the different stages. We ended up spending a lot of time designing fixtures. The design of this in many ways wasn't the design of a physical thing, it was figuring out a process.

Ive looked focused as he spoke. He grabbed different sheets of aluminum extrusions before him, each showcasing a different stage of the machining process that eventually ended up with the final part that houses the main body of the MacBook Air, shown in Figure 7-7.

[7]Jonathan Ive, "Objectified," PromaisMacnews, www.youtube.com/watch?v=nUHROAtyGIg, YouTube video, 05:50, filmed 2008, posted November 8, 2009.

Figure 7-7. MacBook Air aluminum unibody

What I've alluded to is the fact that in order to create products that are truly innovative, not only does the designer have to consider the user, the designer also has to co-invent the processes and systems for which the product becomes reality.

In the wind turbine example, there is a whole process of manufacturing, transporting, and assembling of the various parts that make the project a reality. Because of the enormous size of each part, specialized trucks and equipment had to be designed and adapted to transport the parts to the construction sites on the mountaintops. These trucks carried turbine blades more than seven times their length. So, to move them up the mountains, the blades had to be carried pointing upward toward the sky, as well as having to be rotatable, so that the enormously long blades could avoid hitting trees along the mountain roads, as the trucks drove up. The length of the blades is depicted in Figure 7-8.

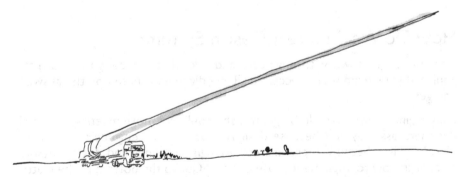

Figure 7-8. Special trucks carrying large wind turbine blades up the mountains of Yunnan Province, China

What these two examples illustrate is the need for designers, especially startup designers who are often pushing to innovate more quickly, to think about the bigger picture in terms of the systems and processes that other members of the team, collaborators, and stakeholders require, in order for the product to be realized. This is the definition of innovation. There are usually no existing processes, systems, or even tools to readily craft something that is groundbreaking. A large part of a designer's job is to account for that.

Systems for Designers and Engineers

As a startup designer, engineers will frequently check in with you to understand how exactly a design is supposed to be executed or when they encounter implementation challenges to your designs.

There might be other designers who are working on parts of the product that complete the parts you're working on. This is when it is appropriate to gather the team and come up with a design system.

Some people in the design community will discuss at length the differences between a design system, a pattern library, and a style guide. However, semantics aside, what the team requires is a shared reference in which different stakeholders can speak the same language when it comes to designing and implementing new and existing features.

A design system typically contains the templates, patterns, components, and, most important, guidelines on how to use them. It is a living document that is subject to change. However, changes must be agreed upon and then publicly communicated to ensure the appropriate updates to existing products if needed.

It is a system that acts as the single source of truth for teams within a startup that design, develop, and realize a product. A good design system helps bring clarity, efficiency, and autonomy to design and engineering teams. A good design system also brings standardization and consistency to the product.

How Do We Create a Design System?

Start by sitting down with engineers and ask what can design do to make your and your team more successful. Then dig deeper, based on the answers you get.

If an engineer says it would be great if she could get more meeting time with designers, ask why. Is it because designs shown to the engineers are incomplete? Or is it because there are not enough guidelines and documentation to explain how to implement those designs? Get to the bottom of the matter.

Then ask yourself if are there things that you're doing that can be templatized, modularized, or automated. Are there things that the team is creating and re-creating for each project? Dig deeper and come up with a list of tasks that could benefit from the simple act for using a template.

A typical design system may look like that illustrated in Figure 7-9.

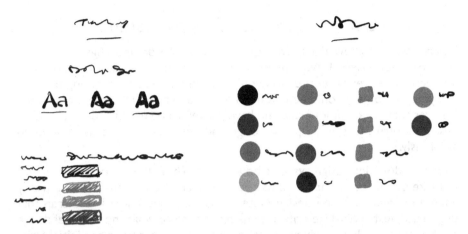

Figure 7-9. Design system

If the product contains a mostly visual user interface, it's worthwhile to consider whether rules should exist for the following aspects:

- Copywriting, voice, and tone
- Typography
- Spacing and grids
- Color
- Iconography
- Illustrations
- Photography
- Motion and animation

Don't design systems for the sake of designing systems, because chances are they'll end up not being applicable. Remember: These templates are meant to be used in real situations and contexts, to help and guide users. Let the user's goals, needs, and context guide your designs, and let your designs guide your systems.

Over time, as the same designs are turned into code repeatedly, it is advisable that engineers also modularize the code they've written in the same format as the design system specifies. In so doing, the design system truly becomes a living, breathing entity that can drastically increase engineering efficiency.

When to Use and Not Use Design Systems

Design systems allow designers not to waste time doing things that in the end don't help users. If buttons in your product are tried and true, there is no need to reinvent them (unless you're in the business of inventing buttons, and your customers could benefit from newer and more innovative ones). For features that are commonplace and have proven to be useful and usable, apply design patterns, to speed up the design and engineer process, as mentioned previously.

However, don't let design systems be a crutch. They should not be an excuse to make designers lazy and lead to cookie-cutter solutions that are arbitrarily limited by what is in the design system toolbox. This is a risk and trap that designers must avoid. Use critical thinking to analyze what the user needs are at each stage of their journey. Forget about design systems at the initial stage of the design process. Instead, think of the ideal solutions first, then let the need for new systems and processes emerge over time.

Summary

Scaling a product is probably the most demanding aspect of being a startup product designer, because there are so many traps and pitfalls along the way. By falling in love with solving the user's problem, never deviating from that core mission, and avoiding the common pitfall of confusing aesthetics with design itself, designers can inspire their team to stay level-headed, even in the midst of fast growth. Also, by designing for accessibility and inclusion through-out the process of scaling a product, the designer can help to ensure that the product lives up to its potential and benefits as many people as possible. Most important, to truly innovate a product beyond the existing cookie-cutter molds, the designer must consider the (new) systems and processes on which the product can be realized.

Psychology, Culture, and Design

Things About the Human Mind and Society That All Product Designers Should Know

Product design is about understanding people and their needs, wants, challenges, and pain points and applying that understanding to create products that address those identified needs, wants, challenges, and pain points. To do so requires a deep understanding of what makes humans human—our thinking, emotions, psyche—what makes us tick, and what motivates us.

One thing is for sure, people will always find ways to use the products we design in ways we didn't expect. That is okay. That is good. Our products should empower humans to accomplish their tasks in every way possible. Therefore, understanding humans doesn't mean we can or should predict their behaviors.

What is important is that we understand as much as possible as we can about human emotions, behaviors, and motivations, so that we can be as considerate as possible and account for as many user situations as possible. This also

© Tony Jing 2018
T. Jing, *Hacking Product Design*, https://doi.org/10.1007/978-1-4842-3985-8_8

doesn't mean that designers act paternalistically and design things that are contrived, obstructive, and allow no room for users' interpretations and applications. There exists a fine balance between being overly presumptive and overly hands-off. A balance of the two is usually an implicit design goal.

Being considerate simply means that we should create products that address real situations and are adaptable and extendable to possible changes in the user's contexts. In this chapter, I will discuss some of the universal principles, theories, and concepts in the psychology of design. Understanding these will help designers to design better and more human-centered products.

Visual Perception

Visual perception is the first and the most common way people are introduced to new products. It has been said that half of our brains is dedicated to seeing and making sense of what we see. This section covers how we view and make sense of things.

Affordance

Affordance simply means that people have an expectation of how something will behave, based on what it looks like. For example, a coffee mug has a handle that invites a person to pick it up by curling his/her fingers inside the handle loop, as shown in Figure 8-1.

Figure 8-1. The handle of a coffee or tea mug is a common example of affordance

Humans naturally understand relationships on seeing these affordances. However, as Don Norman[1] points out in his book *The Design of Everyday Things*, our natural affordances are often met with resistance and challenged as a result of poor design—design that doesn't account for our perceived affordances, a term that Norman coined.

Norman famously gave the example of a door handle that invites users to pull it, only for them to find out that the door requires a push. As a result, the door required a push sign. Even with the sign, however, most people still pulled it at first, as shown in Figure 8-2.

Figure 8-2. Two door designs with drastically different affordances

When applied to designing digital products, affordance means having the right expectations, based on the visual cues presented on screen. More specifically, interactive elements should look like interactive ones. One good example of this is the phenomenon of skeuomorphic design.

According to the Interaction Design Foundation,[2] skeuomorphism "is a term most often used in graphical user interface design to describe interface objects that mimic their real-world counterparts in how they appear and/or how the user can interact with them. A well-known example is the recycle bin icon used for discarding files. Skeuomorphism makes interface objects familiar to users by using concepts they recognize."

[1] Donald A. Norman, *The Design of Everyday Things* (New York: Basic Books, 2013).
[2] Interaction Design Foundation, "What Is Skeuomorphism?" www.interaction-design.
org/literature/topics/skeuomorphism.

Another common example is the design of digital buttons that resemble real-world buttons, with highlights, shadows, and colors that give visual cues similar to their real-world counterparts. When pressed, many of these buttons further mimic the visual effect of a physical button being depressed, with changes in highlights, shadows, and fill colors.

For designers, the understanding and deployment of affordances are key tenets driving the actions of users.

Gestalt

As applied in the fields of psychology and visual design, the word *gestalt* simply means "a unified whole." To understand gestalt is to understand the theories of how humans group visual objects in order to make sense of the world around them. There are six main tenets governing the concept of gestalt: proximity, similarity, continuity, closure, and figure and ground.

Proximity can be explained by this simple rule: objects that are closer together are perceived to be grouped together. A good example of this can be seen in the positioning of photographs and their captions, whether on a web site or a printed page. Even if the objects are identical, proximity alone can create very obvious groupings, as shown in Figure 8-3.

Figure 8-3. The two clusters of boxes appear as two groups

Similarity refers to the human tendency to group similar looking objects with one another and to our ability to pick out repetitions of shapes. As illustrated in Figure 8-4, although the objects depicted are evenly spaced on both the horizontal and vertical axes, and they all share the same color, the rows are perceived as one group rather than as columns.

Figure 8-4. Objects are grouped by similarity

Continuity, on the other hand, is about the perceived continuation from one object to another. As our eyes move from one element or group of elements to another, we follow the path of least resistance. The most obvious example of continuity can be found in implied lines—lines that don't exist but are imagined to exist, as shown in Figure 8-5.

Figure 8-5. Lines don't have to be continuous

Closure is similar to the concept of continuity, in that our brain is completing a shape or an outline, based on the space created by objects nearby. As the name suggests, our brains are making a jump to create closure by joining spaces together. Both continuity and proximity influence closure, as shown in Figure 8-6.

Figure 8-6. A square shape is formed in space

Figure and ground denotes the way in which our eyes differentiate an element or an object from its surrounding environment or space. The relationship between figure and ground is spatial. Figure is usually thought of having shapes, while the ground is usually not. As shown in Figure 8-7, the black dots are usually perceived as the figure, and the white surrounding it as the ground.

Figure 8-7. Black dots appear as the foreground

Comprehension and Memory

The first thing to note is that reading and comprehension are different things. The context of the reader has an enormous impact the reader's comprehension and recall. Reading by no means guarantees either of the two.

The second thing to note is that memory is always fuzzy, and that for information to be retained, people have to proactively use it. The point about memory being fuzzy points to how we reconstruct memory each time we bring it back in our minds. Rather than brute force memorization, it is easier to give people cues, so that they can recall information.

Another reason to design for recall is because people will forget things. Memory fades over time. Therefore, when designing, we should not rely on people remembering things. We should anticipate when certain information becomes pertinent and present it to users in a timely fashion and create easy ways for them to look things up.

Progressive Disclosure

The concept of only providing information that is the most relevant at a time is called progressive disclosure. Ensuring this relies on separating a large chunk of information into layers that are presented individually, often in sequential order.

A good example of this can be seen in the sign-up flow for Oscar Health, a new kind of health insurance company. Rather than showing a page of a dozen questions that could overwhelm the user, Oscar Health shows one question at a time, allowing the user to focus on answering each question, without being distracted by how many questions there are in total.

When a complex procedure is a required path for users to complete a task, use progressive disclosure to guide them through the process.

Mental Models

Mental models are the representations of systems and objects in people's minds that are conjured, based on their past experiences. People can form models of how these things work and of how they can be interacted with. An example of a mental model can be seen in the emergence of online subscription and mobile services. Newspapers and magazines had already established the subscription model of paying a repeating fee periodically to access publications. Back in the early 2000s, the predominant model of selling software was still hard-copy based, meaning that each piece of software was bought and sold on a CD or a hardware package that was either taken home by the buyer from a store or mailed to his/her address.

With advent of the Internet, buying and selling gradually moved online, with the software being downloaded onto buyers' computers. However, as time moved forward, many businesses shifted away from this model, offering their software as a service (SaaS), allowing users to subscribe to their software products, just as they subscribed to magazines and newspapers.

The use of the word *subscription* helped to establish this clear mental model in people's minds. Users knew how this system would work: they pay a monthly or yearly fee, in exchange for use of the software, as if they owned it.

In the print world, the act of subscribing to a magazine meant either calling or mailing the magazine's sales office with a request to subscribe, a method of payment, and the intended mail delivery address. It was not immediately clear how the interaction model of this new type of online subscription would work. Therefore, it was the designer's job to create new ways of interaction that were easily learnable and could form a bridge between the old mental model and the new model of the system.

Through this process of teaching new interactions and bridging the old mental model to new systems, new mental models can be created. Today, "online subscription" has become a well-established mental model in itself.

Metaphors, Examples, and Stories

In some sense, both mental models and affordances can be considered to be metaphors. Both are about one set of expectations being transferred to another similar situation, allowing the user to rely on previous experiences to make sense of new situations.

The use of examples is an effective way to spread ideas about a topic. Products that involve complex processes often give extensive demos or usage examples, illustrated in a learning tutorial. This is why complex and usually expensive products, such cars and machinery, are often presented with either real-life demonstrations or samples, or with comprehensive training videos, in which the user is shown how to use the product.

Both metaphors and examples assume that people are interested in learning about the information presented. However, the best way to convey information at a subconscious level is through the use of stories. Before there were written words, human beings passed down stories by memorizing them into rhyming verses. This is the original way in which knowledge was passed. Maybe that's why stories are so powerful at communicating information; our attention is almost instinctively drawn to them.

When it comes to product design, a good use of stories can make information not only interesting but also understandable and memorable for the long term.

A story can take many shapes and forms. It can be oral, visual, textual, or told through the use of sound and moving images. However, regardless of its shape or form, a story usually involves a few of the following aspects:

- Context
- Characters
- Plot
- Mood

Context gives the story its setting, in terms of time and place, while characters make the story relevant, by offering protagonists and antagonists, in addition to their relationships provided by the context. The plot moves the story along and ties the characters and context together. The mood is the emotional tone of the story, set by how the plot reveals itself, along with the stylistic application of language, imagery, and music, to complement the plot revelations.

Most stories take on a chronological narrative and imply a causal relationship of events. Take advantage of this natural tendency in people to make your points of view more interesting. Especially when information is dry and verbose, storytelling can greatly improve the experience of engagement and recall of the information conveyed.

When applied to experience design, the users' emotional response to storytelling further enforces their ability to recall information and learning, because emotional engagement makes a story personal and more easily internalized.

Culture

Culture is defined as "the customary beliefs, social forms, and material traits of a racial, religious, or social group; or the characteristic features of everyday existence (such as diversions or a way of life) shared by people in a place or time," according to the Merriam-Webster dictionary.[3] Simply put, culture can be both the norms and social behaviors formed within a group of people, and the group's shared social characteristics.

Archetype

There are almost uncountable numbers of cultural groups and subgroups. People who live within a national boundary are often in the same national cultural group. People who share similar interests in paper crafts are in the same interest-based cultural group. With the emergence of digital communication, groups don't necessarily have to be divided by the boundaries of geography.

Common themes of motifs can be found in these cultural groups. Over time, these themes may develop into archetypes, which are the manifestation of said themes in literature and imagery.

The fashion and clothing brand Brooks Brothers adopts archetypes in the design of its products. A keyword-association exercise of the brand would commonly bring forth such words as *luxury, statesmen, American,* and *traditional*. The high-end, historic statesmen myth is used as an archetype, which

---

[3]Merriam-Webster OnLine, s.v. "culture," www.merriam-webster.com/dictionary/culture, July 1, 2018.

drives the look and feel of its products and its brand. Unlike Brooks Brothers, GoPro, an action camera company, aligns itself with a totally different archetype. By using extreme sports athletes who personally use the camera to record their own adventures as promoters of the brand, GoPro evokes the hero archetype. In fact, its flagship camera is named GoPro Hero. GoPro even sources action videos shot on GoPro cameras by its customers and features them on its YouTube channel. Brooks Brothers, however, does not invite its customers to document and share their experiences, because its archetype is more subdued and reserved. The fact that Brooks Brothers outfitted 40 past US presidents is enough to drive home its archetype. Having extreme sports athletes promote Brooks Brothers would be far less effective.

The key here is that much of product design starts with understanding the common archetypes that match the product. Applying and aligning the right archetype with the product is a significant factor in a product's long-term success.

Design Reflects Culture

Zooming out a bit further from archetypes, we'll realize that all design reflects the broader cultural contexts of its surroundings, whether physical or virtual. A good example of that reflection can be seen when products designed for the US market and Japanese market are compared.

Quite often, when Americans travel to Japan, they'll realize that many day-to-day products are often about 10% to 20% smaller in Japan. A very visible example of this would be the size of automobiles. American cars are huge compared to Japanese ones (the ones made for the Japanese market). While America's vast landscape and relatively sparsely distributed population affords it the ability to enjoy such vehicles as pickup trucks, Japan is a totally different story.

As soon as you enter a Japanese city or town, you'll see smaller vehicles on the road. These small cars, trucks, and vans are known as *Kei cars*, translated literally as "light automobiles." Not all Japanese vehicles are these small Kei cars, but a significant portion of them are. Perhaps, on first glance, you might not notice the difference, but when compared with other vehicles, as shown in Figure 8-8, the differences become obvious.

Figure 8-8. A Kei car compared with a standard American sedan

It's obvious that the need for Kei cars has something to do with the width of Japanese side streets in particular. In most Japanese towns, the side streets are often no wider than 3 or 4 meters (10 to 13 feet), much smaller than typical American neighborhood streets, thus imposing limits on the size of cars and trucks.

An effect of cultural difference on thinking is highlighted in Susan Weinschenk's book[4] *100 Things Every Designer Needs to Know About People*, in which a research project comparing people's visual attention observed that, as shown in Figure 8-9, "if you show people from the West a picture, they focus on a main or dominant foreground object, while people from East Asia pay more attention to context and background. East Asian people who grow up in the West show the Western pattern, not the Asian pattern, thereby implying that it's culture, not genetics, that accounts for the differences."

---

[4]Susan Weinschenk, *100 Things Every Designer Needs to Know About People* (Berkeley, CA: New Riders, 2011).

Figure 8-9. People were asked this question: What do you notice more: the cows or the background?

The theory highlighted in the research is that the cultural norms of East Asia are more focused on relationships and groups, as compared to Western norms. In the West, individualism has slightly more focus; therefore, attention is paid to focal objects.

China, a neighbor of Japan, is also situated in East Asia and has its own densely populated cities and tight neighborhoods. So why doesn't China have small cars, as in Japan? It could be because of three things. First, it's the transforming landscape of China. While Japan has preserved much of the neighborhood road structures from its past, nearly a century of war, revolution, and recent construction projects have changed China's cities drastically. Now, neighborhood roads are much wider in cities than centuries ago.

Second, Japan industrialized earlier than China. While China has been catching up rapidly, there is still a long way to go when it comes to automobiles per capita. For every 1000 people, there are 591 cars in Japan, whereas there are only 83 in China.[5] This means that while the richer Chinese cities are already filled with all kinds of cars, in the countryside, where narrower village roads still exist, the phenomenon of an automobile in every household has yet to appear.

The third reason why small cars likely wouldn't take off in China might have something to do with how culture influences values and thinking.

As do Americans, the Chinese prefer larger things in general. The Chinese concept of *dada fangfang* ("big and generous") can be seen in the design of

[5]NationMaster, "All Countries Compared for Transport Road Motor Vehicles per 1000 People," NationMaster.com, www.nationmaster.com/country-info/stats/Transport/Road/Motor-vehicles-per-1000-people, citing Wikipedia, "List of countries by vehicles per capita," https://en.wikipedia.org/wiki/List_of_countries_by_vehicles_per_capita, updated August 18, 2018.

chopsticks (see Figure 8-10). Chinese chopsticks are evenly proportioned and longer than the other two pairs illustrated. The Japanese have no such preference; instead, the concept of *miyabi* ("elegance") is emphasized. This might be why Japanese chopsticks are usually shorter, with a pointy end.

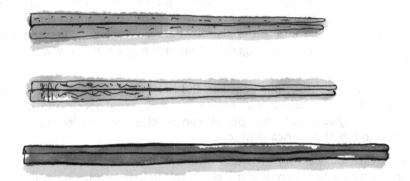

Figure 8-10. Japanese, Korean, and Chinese chopsticks

This preference for aesthetics can have a big impact in architecture as well. While the Kyoto Imperial Palace is known for its elegance, the Forbidden City in Beijing is renowned for its scale and grandeur (see Figure 8-11).

Figure 8-11. The elegant Kyoto Imperial Palace(left) and the imposing Forbidden City in Beijing

Railway stations also reflect this pattern. Both Tokyo Station and Shanghai Hongqiao Railway Station are major transportation hubs connecting multiple railways, airports, and subways. Both move hundreds of thousands of passengers every day, with a high frequency of bullet trains running at 300 km/hour (190 mph) arriving and departing on the dot.

While each complex has an impressive facade, things differ on the inside. Tokyo Station has hardly any centralized waiting areas for the multiple train routes, but that works fine. People move about freely in the complex, hopping on and off trains. There are shops, restaurants, and vending machines nested throughout the complex, if passengers choose to stop by.

In Shanghai, rather than placing waiting areas along the different platforms, passengers wait for trains in a centralized grand hall above the platforms. All of the shops and dining areas are aggregated in the corridors to the sides of the waiting hall. The doors to the platforms are only opened when the trains approach, at which time passengers line up in the grand hall and walk down to the platform.

Our environment shapes our behavior, and our behavior shapes our culture. Our culture, in turn, affects our environment. Personal space (the distance from one's body to which other people can get close without feeling invasive) is greater in North America than in Asia.

Times Square can be very crowded on New Year's Eve, but that density of people is actually quite common in Asia. People in East Asia are generally more accustomed crowds and being part of one.

Figures 8-12, 8-13, and 8-14 compare three news aggregator sites (in traffic) from the United States, Japan, and China (excluding search engines, social media/networks, and e-commerce sites).

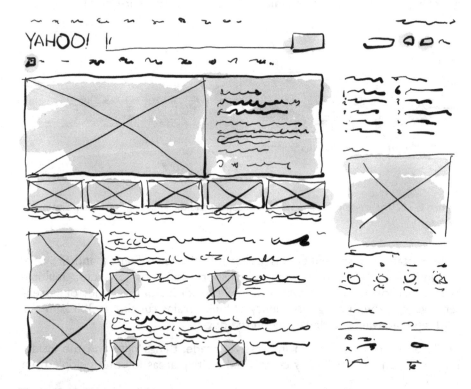

Figure 8-12. Yahoo.com's layout

Figure 8-13. Yahoo.co.jp's layout, which is narrower than Yahoo.com's

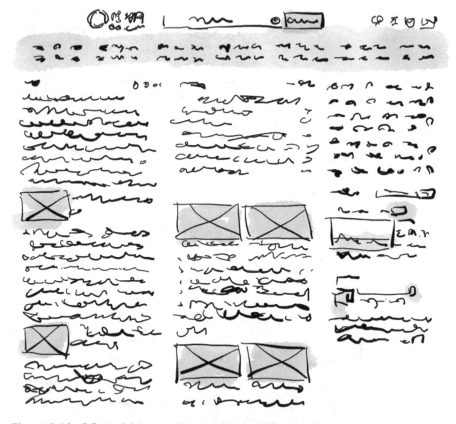

Figure 8-14. QQ.com's layout is slightly wider than Yahoo.co.jp's

Both Yahoo Japan and QQ.com show more content than Yahoo.com. This is because logograms, by their nature, contain more meaning. However, a few subtle differences can be observed between QQ.com and Yahoo Japan. QQ.com is slightly wider and has a slightly bigger character size.

One thing to learn from recognizing culture's impact on design is that designers should understand the relationship between the environment, the behaviors, and the cultures that surround the products, systems, and processes that are to be designed for them.

This is because cultural forces that influence each single member of a certain culture inevitably influence the output of design work. In turn, the designed products also affect the culture. In this sense, designers should be very aware of both the greater culture they operate within and the cultures for which they're designing.

Summary

Psychology and culture both shape our behaviors, mindsets, and expectations. They help us to make sense of our physical and metaphysical environments. In order to design products for people, designers must understand people first, especially their psychology and culture.

Principles such as affordance and gestalt can be applied to design more intuitive and usable products, whereas an understanding and use of progressive disclosure, mental models, metaphors, and archetypes allow designs to influence user behavior. Last, design always exists in the context of culture. As product designers, we must be aware of the impact that culture and subcultures have on our products.

Tools, Frameworks, and the Future

This chapter is about the tools and frameworks product designers can use to strengthen the quality of their work and evaluate both their work and that of their peers. The latter part of this chapter is about the common pitfalls that product designers might face and the ways to avoid them.

The first thing I should clarify is the definition of tools. When discussing design tools, most designers think of design software programs, such as Sketch and Figma, or prototyping programs, such as Principle or Framer. These are not what is being referred to in this chapter. The tools we are concerned with are more akin to ways of doing things, rather than physical objects or digital software. Although these methods and ways of working are somewhat abstract and fluid in their application, they have proved to provide immense value in ensuring the effectiveness of design.

A Recap

Let's start with a quick recap of the different frameworks and ways of working covered in the previous chapters.

© Tony Jing 2018

T. Jing, *Hacking Product Design*, https://doi.org/10.1007/978-1-4842-3985-8_9

Useful, Usable, Feasible, Viable, and Desirable

In Chapter 2, I discussed that the ultimate hallmark of a great product is its ability to meet different and often competing goals: usefulness, usability, feasibility, viability, and desirability.

A product should first and foremost be useful. This means that it should help people solve a problem or do a job. Beyond that, it should also be usable, meaning that the product should be intuitive and easy to learn. It doesn't matter how useful the product is, if people can't figure out how to use it. The usefulness and usability of a product are attributes measured from the user's perspective.

On the other hand, feasibility and viability are attributes measured from the business's objective. So despite the tremendous business potential (viability) of a time machine, such a product is simply not feasible, because the required technology simply does not exist today. The viability of a product is its ability to generate long-term success. This has to do with a product's ability to scale and contribute to profitability over time.

In the late 1990s, technology for small portable music players became feasible. Companies such as Apple capitalized on this opportunity and created MP3 players that were quickly adopted by millions around the world (Figure 9-1).

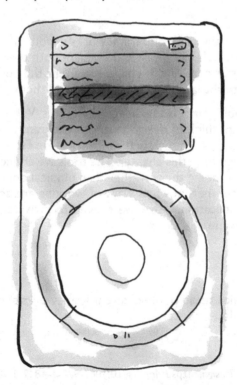

Figure 9-1. Apple's iPod

Apple's iPod is a great example of feasibility enabling viability through desirability, because Apple innovated against its feasibility limits and created a device that was not only designed well but also functioned well, and the product was marketed well, so it became a business success. By the mid-2000s, the iPod had Apple doubling, tripling, and quadrupling its earnings over a number of years.[1]

Two Goods

In addition to being useful, usable, feasible, viable, and desirable, two more attributes should be considered in the evaluation of a product's success. They are: (1) the product should be good for the immediate society it serves; and (2) it should be good for the world in general.

In the case of the iPod, while it created hundreds of millions of hours of enjoyment, a bigger issue loomed over it. As more and more people bought electronic devices in the early 2000s, electronic waste became a real problem for many developing communities left dealing with the storing and recycling of these wastes. While Apple was not the worst offender in contributing to this problem, it certainly helped to kick-start the craze over electronic devices and the trend of frequently replacing old devices with the new ones annually. Another example of this is Facebook, the global social networking juggernaut. While Facebook connected billions of people around the world, its vulnerability in being hijacked to affect the democratic process emerged after the 2016 US presidential election. In this sense, it can be argued that while on the whole Facebook has provided benefits to billions of users, its net effect on the world is not as positive as was believed.

The request here is not that product designers foresee all potential risks and pitfalls. That is simply impossible. The request is that we, as professionals, begin to think about the possible consequences of the work we do. While certain things will always surprise us, it helps that product designers think about whether the solutions they propose are good for society and the world in the long term.

Values, 10x Better, and Details

In Chapter 2, I also discussed the importance of taking a stand about the kind of future that we would like to see. This vision should come from the company's mission statement and distinct values.

---

[1]Katherine Griffiths, "Apple Profits Quadruple on iPod Surge," *The Independent*, www.independent.co.uk/news/business/news/apple-profits-quadruple-on-ipod-surge-486472.html, January 13, 2005.

The product designer should work with the product manager and engineers to create a product that is ten times better than the incumbents' in at least one particular experience or flow. The product should have enough of the important details right, so that it becomes a sticky product—one people want to keep buying.

Jobs to Be Done

In Chapter 3, I discussed using the jobs-to-be-done model to figure out what features to build, so that the product helps people in concrete ways. The jobs-to-be-done model scan be distilled into one sentence:

> *When ___, I want to ___, so I can ___.*

With this sentence, customers provide a map of their needs, goals, and the exact jobs they are trying to "hire" products to do. The jobs-to-be-done model helps the product team to translate customer's goals into tangible features.

Narrow and Deep

There are billions of jobs that must be done in the world. How do we determine which ones we should design for first? Chapter 4 shows us how to find the answer.

The key is to focus and choose to work on specific kinds of jobs—ones that are narrow and deep, meaning that they should fall within specific markets and industries that the company aims to tackle. The goal for the product team should be to get a small group of people who share the same job to become really excited about the potential solutions you're providing. Even if the number of people in that group is small, as long as your solution provides a significant positive impact on their jobs, the product has a good chance of growing into something much bigger over time.

Build, Measure, and Learn

How do we know if our product has hit the mark or not? The answer is that we won't know until our customers show us. This is why we should follow the iterative product development approach, as highlighted in Chapter 4, which is that we build first, measure the effects on what we build next, and last, learn from what we have observed. What makes this process iterative is that we take the learning to improve on the product we began with and build the next round to be a lot better.

The Only Metric and 80/20

In Chapter 6, we dug in further and revealed that when it comes to designing an MVP, the team should identify a single metric as the key indicator of success. Having this metric will help the team focus and get to a baseline level of success, in terms of proving the initial hypothesis.

Through this lens, product designers should adopt the 80/20 rule when it comes to their day-to-day work. They should seek to understand what key levers (the 20%) of the product are driving the majority of positive output (the 80%), whether growth, profitability, or hours spent.

Altogether, the team should first identify a job to be done in a narrow and deep problem space. Then, the team should iterate the product by building, measuring, and learning against a single metric. As use of the product grows, and the hypothesis adapts, based on user feedback, the team should repeat this process iteratively, to increase the problem/job scope of the product.

Space-Time Continuum

A good framework to use when it comes to evaluating the quality of existing experiences is the use of a space, time, emotion, and value continuum. Some people call this continuum a journey map or a user experience map. In essence, it is a chart documenting the user's experience with a specific product or service over the course of her/his involvement with the product or service.

Traditional journey maps come in many shapes and forms, but good ones generally share the following traits. First, the information on the maps is always layered. The first layer of information is the passage of time indicated by user tasks. The second layer is usually the touch points the user faces. The next layer is usually the user's emotions. Figure 9-2 is a sample journey map of a retail shopping experience, with the first three stages shown.

Stages	**Awareness**	**Discovery**	**Entry**
Touch Points	Advertising online and on TV, billboard on the highway	Storefront visible on the street	Store clerk greetings
Emotion	Calm	Intrigued, captivated	Overwhelmed, delighted

Figure 9-2. Journey map of a visit to a retail store in a busy part of town

Journey maps are the visual documentation of customer experience. They should be easily understandable and shareable. They can be a powerful artifact that creates a shared understanding of how customers and users discover, interact, or even reject your product. By creating this map and coming to an agreement about the assertions made by it, the product team can not only feel empathy for the users but also understand where the product does its job and where it lags. The key here is that the assertions made by the map must be backed up by real use cases and observations. The moment we start guessing and conjuring ideas about what the user might feel or experience is the moment the chart loses its usefulness. An accurate account of the experience has to be reflected.

While the preceding journey map works at a baseline level, it can be improved. First, the user's goals at each stage have not been clearly identified. Also not considered is the user's physical environment. Depending on the context of the product, the user's goals and the user's physical environment could change significantly as time passes. Without considering these factors, it is much harder to holistically evaluate and pinpoint the reasons the product or service is able to succeed. With this in mind, I recommend adding two more rows: user goals and space.

Rather than calling it a map, a more appropriate name is a matrix, because it is a table evaluating the interplay of different parts of the flow (the rows) and factors that contribute to the user's changes in goals and emotions (the columns) (see Figure 9-3).

Stages	**Awareness**	**Discovery**	**Entry**
Touch Points	Advertising online and on TV, billboard on the highway	Storefront visible on the street	Store clerk greetings
Space	At home, at work, on the computer, on the phone, driving	Walking on crowded sidewalk	The store is bright There is a strong aroma
Emotion	Calm	Intrigued, captivated	Overwhelmed, delighted
User Goals	"Dont interrupt me."	"Help me understand what you sell."	"I want to find the products I'm interested in quicker."
Problem/ Opportunity	Lack of awareness about online store	Storefront is very busy	Too many strong senses

Figure 9-3. The space, time, and experience matrix

Problem/opportunity is an additional row that can be added to this matrix, especially in the scope definition and problem-solving stages of the project. Once the problems have been identified and agreed upon, this addition allows the team to shift into thinking about ways of improving. A problem/opportunity list at this stage helps to keep the forward momentum.

Cover Your Bases

The beauty of the space, time, and experience matrix is its versatility and universality. It can be applied to all kinds of experiences and products, in almost all situations and contexts. However, beyond its application in evaluating an existing experience, it can be also be applied in the design process, to evaluate the effectiveness and completeness of a proposed design.

In particular, it works well as a follow-up to the jobs-to-be-done model, because it allows the specific flow and features of the job to be mapped. By putting this matrix of the proposed flow and desired customer experience in front of the team, the design can be evaluated holistically, and potential gaps can be uncovered.

Happy Path

There are a few specific ways to apply the matrix for that evaluation. The first is to examine the happy path. The happy path is, almost by default, the main flow that designers come up with. We all hope our products provide the ideal experience, and as a result, we design for that ideal experience first.

Adhithya Ramakumar, a designer working at Google, pointed out that the design of the happy path should involve the designer accounting for different scenarios.[2] These scenarios are states and conditions of the product experience that are significant enough to cause major variations in its flow. The key here is to cover all those cases and account for each fork in the road. This is the design of the happy path.

While in general it is better to provide a good experience for users when they use your product, the most crucial aspect of them using your product is that it does the job it is supposed to do. This is the priority. As long as the product gets the job done, and the experience of using the product doesn't get in the way of that objective, users won't care if your path isn't perfectly happy. Of course, we should always strive to provide ideal experiences. However, it is paramount that we know the user's priorities and get the job done.

Edge Cases and Unhappy Path

What are the possible reasons your product won't get the job done? This is the main question that we must ponder and answer in the design process. Write down a list of potential answers by brainstorming and plotting them inside the space, time, and experience matrix. Are these flows the best they can be? If the answer is no, our work is not done.

Many junior designers will stop at the creation of a happy path and consider their work done. However, the true hallmark of a good product is not how well its happy path performs, it is how well it handles edge cases and error states. Although we should do our best to avoid them, edge cases and errors

---

[2]Adhithya Ramakumar, "An Interaction Design Framework to Cover All Your Bases," UX Collective, https://uxdesign.cc/a-interaction-design-framework-to-cover-all-your-bases-1e616a954827, May 08, 2018.

will always exist, because humans make products, and humans are prone to making mistakes.

Because of those risks and external events beyond our control, our product always has a chance to fail our users. In such cases, we should still fail gracefully, by giving users as much peace of mind as possible.

The first step to failing gracefully is to account for as many error messages as possible and make sure the error messages are understandable and contextual. Be sure to converse with users in a polite and jargon-free manner. Always provide an explicit action to resolve the error or a recommendation for a next step, if the problem cannot be resolved quickly.

Audit the Product's Touch Points

All products communicate with their users. When a product stops communicating, it stops being useful. We can tell whether a light fixture is on or off by simply looking at it. In this case, the product instantly communicates its state to us. This is the simplest form of communication. It's direct and visual.

This sounds obvious, but as products grow in complexity, the complexities by which these products communicate with their users grow as well. A simple visual inspection usually will not work to gauge the status and conditions of a large enterprise's software.

Clarity and purpose are the two attributes of good communication. A good principle to adhere to is that all communication should be explicit and unambiguous. This is especially true for user actions that cause change within a system. For actions that are critical or permanent, allow users to confirm their decision by using a *two-step operation*. The iPhone's turning off device flow is a two-step operation. It first requires that the user press the power button for a relatively long interval, then swipe over the digital slider to the right. By dividing the switch-off feature into two parts and using a slider interaction to confirm the switching off, the risk of accidentally turning off the phone is greatly reduced.

Furthermore, for actions that are error-prone, make them reversible. A good example of this is Gmail's temporary undo button, which appears after sending an e-mail. Gmail surfaces this button for a few seconds, just in case the e-mail has been sent by accident.

An often-overlooked part of the design process is communication with the user outside of the core experience. This communication can take the form of text messages, e-mails, phone calls, or even snail mail. Always validate the flow with a space, time, and experience matrix and design the right messaging, regardless of the medium or place within the experience.

Communication and Critique

Good communication is a prerequisite for good design. Whether it's a review meeting with the product manager and the engineers or a meeting with a prospective client, a business partner, or an investor, the startup product designer is often expected to present and explain his or her work. Here are a few steps to follow to make the best out of these meetings.

The first thing to do is to explain the context and set the background for the problem, in terms of the user's pain points, goals, and jobs to be done. Then set the expectation of how much material the presentation will cover. A three-step flow requires that much more time be spent on each step than a twenty-step flow.

Begin by explaining the goals of the design, then explicitly state what kind of feedback you're looking for, whether it's the business potential, the product flow, the UX or the UI, etc. Create a framework for the audience to first understand and then evaluate the design. Show the space, time, and experience matrix, to help to establish that framework and get the audience to think holistically.

If the audience consists of designers or product people, the design process could also be discussed. If not, the designer should focus on the opportunities explored and the landed solution. After analyzing the solution, the designer can provide the potential next steps and ask for feedback.

The key is to let the feedback land in the meeting and analyze it after. The reason for this is that the description and recommendation in the feedback should be given the opportunity to be heard, in order for it to be evaluated. It is important, however, not to confuse hearing feedback with committing to what's discussed right then and there. Designers have to hold firm on this.

This is because we should evaluate feedback and filter the bad from the good. Common examples of bad feedback include a focus on personal preferences, those that don't take account of the context and the problem. Another type of bad feedback is that directed at the designer rather than the design. Beware of logical fallacies, such as appeal to authority—the belief of something being true because an authority figures claims it to be true—and of the straw man, an idea that is misrepresented, so that it can be more easily refuted.

At the time of the presentation, however, listen carefully and write things down. Better yet, ask your participants to write down their feedback. Ask for clarifications and fully recognize where the feedback is coming from. Doing so will make later analysis and evaluation easier.

The Future

In his talk titled "The Future Mundane," Nick Foster,[3] head of design at Google X, spoke about the tendency of designers to design mainly for a hero figure or "a trickled-down aspirational super user intended to give us all something to hope for," when it comes to designing for the future.[4] Foster argues that instead of designing for those characters, we should design for the common people, the ones in the background who are neither heroes nor villains, because our world is not filled with heroes or villains but rather people like you and me.

I think his argument touches the core of what it means to be a startup designer, which is that, fundamentally, all problems are worth solving, and it is up to us to uncover and solve them.

Being in a startup means designing products that start small, initially solving problems for a narrow audience but doing our best to help it. This means that, as product designers, we should have the humility and empathy to understand people—their challenges, goals, and values—then use our knowledge of human cognition, behavior, and culture, in addition to our skills in crafting interfaces and experiences, to truly make a positive difference in people's lives.

Carry On Designing

This concludes the book. The skills, tools, guidelines, and heuristics covered in it are to be applied by you. They don't exist in a vacuum. Whether they should be applied depends entirely on your understanding and evaluation of the context. So, use critical thinking to assess each situation, and make your own judgments.

Happy designing!

---

[3]Nick Foster, "The Future Mundane," Vimeo, https://vimeo.com/139358108, September 15, 2015.
[4]Nick Foster, "The Future Mundane," Core77, www.core77.com/posts/25678/the-future-mundane-25678, October 7, 2018.

Index

© Tony Jing 2018
T. Jing, *Hacking Product Design*, https://doi.org/10.1007/978-1-4842-3985-8

Printed in the United States
By Bookmasters